Visual Mental Imagery System: How We Image the World

Visual Mental Imagery System: How We Image the World

Guest Editor

David F. Marks

MDPI

Basel • Beijing • Wuhan • Barcelona • Belgrade • Novi Sad • Cluj • Manchester

Guest Editor
David F. Marks
Independent Researcher
Arles
France

Editorial Office
MDPI AG
Grosspeteranlage 5
4052 Basel, Switzerland

This is a reprint of the Special Issue, published open access by the journal *Vision* (ISSN 2411-5150), freely accessible at: https://www.mdpi.com/journal/vision/special_issues/I90201SVB1.

For citation purposes, cite each article independently as indicated on the article page online and as indicated below:

Lastname, A.A.; Lastname, B.B. Article Title. *Journal Name* **Year**, *Volume Number*, Page Range.

ISBN 978-3-7258-4761-7 (Hbk)
ISBN 978-3-7258-4762-4 (PDF)
https://doi.org/10.3390/books978-3-7258-4762-4

Cover image courtesy of David F. Marks

Contents

About the Editor

David F. Marks

Dr. David F. Marks is an independent researcher who is interested in cognitive psychology, consciousness, mental imagery and health psychology. In 1966, David graduated with a BSc in Psychology from the University of Reading, UK. In 1970 he obtained a PhD at the University of Sheffield, UK. From 1970–1986, David worked as a Lecturer/Senior Lecturer at the University of Otago in New Zealand. From 1986–2000, he served as Professor of Psychology at Middlesex University in London, UK. Then from 2000–2010, David served as Professor of Psychology, City University, London, UK. From 2010 to the present date he has served as a journal editor, published several books, and carried out research as an independent scholar.

vision

MDPI

Editorial

Introduction to Special Issue: Visual Mental Imagery System: How We Image the World

David F. Marks

Independent Researcher, 13200 Arles, Provence-Alpes-Côte d'Azur, France; dfmarksphd@gmail.com

The aim of this Special Issue is to provide new perspectives on the role of visual mental imagery in how we image the world, past, present and future. This Special Issue comprises research articles from across the spectrum of relevant empirical sciences. We are interested in the role of visual and mental imagery in cognition, imagination, and action directed towards the present or anticipated states of the external world. This Special Issue was designed to enhance our understanding of how human beings image the world with self-awareness in relation to a challenging and changing external environment. Within this theme, there was scope for studies addressing individual differences, phenomenology, imagination, creativity, artistic expression, forecasting, and futures studies. This Special Issue consists of ten articles.

The first article (Contribution 1), by David F Marks, reviews phenomenological studies in visual–mental imagery (VMI) over early (1860–1929), middle (1930–1999), and recent (2000–2023) time periods. The review focuses on two forms of constructive VMI—memory and eidetic imagery—that vary along a continuum of vividness. Vividness is defined as "a combination of clarity, colourfulness and liveliness, where clarity is defined by brightness and sharpness, colourfulness by image saturation and liveliness by vivacity, animation, feeling, solidity, projection and metamorphosis." The author integrates his findings into a template with 16 vividness properties to reveal a high degree of consistency in the reported VMI characteristics across time, space, and cultures.

The second article, by Nadine Dijkstra (Contribution 2), addresses the question of whether the early visual cortex (EVC) is necessarily involved in VMI. Dijkstra proposes that inconsistencies in findings regarding the EVC in VMI can be understood as being due to "unique challenges associated with investigating EVC activity during imagery." The author suggests that EVC activity relates to specific imagery properties. Methods sensitive to variation in low-level features indicate that imagery is able to recruit the EVC in similar ways to those of perception. However, if the requisite level of detail is absent, then the EVC is unnecessary in VMI.

The third article, by Feiyang Jin, Shen-Mou Hsu, and Yu Li (Contribution 3), comprises a systematic review of search of English-language articles published between 2015 and the present. The authors pinpoint three primary areas of disagreement: whether aphantasics exhibit impairments solely in visual imagery, whether a distinction should be made between the absence of voluntary and involuntary imagery, and whether aphantasia primarily concerns the generation of imagery. The authors provide a cutting-edge review of the conceptual and methodological issues at the core of aphantasia research. They suggest that future research is needed to "refine the definition and diagnosis of aphantasia, strengthen empirical investigations at behavioral and neural levels, and, more importantly, develop or update theories".

The fourth article, by Andrea Blomkvist (Contribution 4), reviews the contributions of the hippocampus to mental imagery formation, especially evidence supporting the view that the hippocampus contributes to the spatial model used for mental imagery. The author argues that different hippocampal circuits contribute to different types of imagery, such as object imagery, scene imagery, and imagery with a temporal aspect. Blomkvist suggests that new avenues should be opened up for research into the role of the hippocampus across a variety of imagery tasks.

The fifth article, by Takao Hatakeyama (Contribution 5), examines empirical associations between imagination, mental imagery, and relevant cognitive abilities and autistic traits. The author hypothesizes that imaginative impairments and distinctive sensory characteristics are to be expected in the mental imagery capabilities of individuals with autism spectrum disorder (ASD). Two vividness tests and one imagery-type test were used to assess imagery ability in 250 college students. A high-scoring ASD group exhibited lower visualization scores, indicating that autism-like traits are associated with mental imagery ability.

The sixth article, by Hiroki Motoyama and Shinsuke Hishitani (Contribution 6), presents a study of the brain regions associated with the perception-driven suppression of mental imagery generation. The focus of the article is the left posterior cingulate gyrus (PCgG). The authors compared brain activation in a picture observation condition and a positive imagery generation condition. Significantly higher activation occurred in the left PCgG in the former condition compared to the latter condition. A significant correlation was observed between the activation of the left PCgG and participants' subjective imagery richness ratings, which are a measure of the clarity of a presented picture. The results suggest that the left PCgG plays a role in suppressing the generation of mental imagery.

The seventh article, by Mohamad El Haj and Ahmed A. Moustafa (Contribution 7), investigates eye movements during near- and distant-future thinking. Previous research suggests that near-future events are typically viewed from a first-person (own-eyes) perspective while distant-future events are typically viewed from a third-person (observer) perspective. Would these distinct mental perspectives be accompanied by distinct eye movement activities? To test this hypothesis, El Haj and Moustafa invited participants to imagine near- and distant-future events while recording their eye movements with eye-tracking glasses. Fewer, longer fixations were observed for near-future thinking than for distant-future thinking, and more "field" mental visual perspective responses were evident for near-future thinking than for distant-future thinking. This study demonstrates the links between the temporality of future thinking triggers and distinctive mental imagery and eye movement patterns.

The eighth article, by Ekaterina Pechenkova, Mary Rachinskaya, Varvara Vasilenko, Olesya Blazhenkova, and Elena Mershina (Contribution 8), reports an investigation into the ability to adopt different perspectives in mental imagery. Pechenkova et al. used fMRI to compare functional connectivity during first-person and third-person perspective taking among 26 participants performing two spatial egocentric imagery tasks, imaginary tennis and house navigation. Although no significant differences were indicated, a "small subnetwork of early visual and posterior temporal areas" yielded stronger functional connectivity during the first-person perspective, suggesting the possibility of an associated "loop between long-term memory and the "visual buffer" circuits."

The ninth article, by Maren Bilzer and Merlin Monzel (Contribution 9), describes a study comparing the multisensory profiles of voluntary mental imagery and dream imagery in 226 participants. Bilzer and Monzel report an association between the vividness of voluntary mental imagery and the vividness of dream imagery that appeared to be moderated by the frequency of dream recall and lucid dreaming. Emotional and visual imagery were significantly more vivid for dream imagery than for waking imagery, while the vividness of auditory, olfactory, gustatory, and tactile imagery was higher for waking mental imagery.

The tenth article (Contribution 10), by David F Marks, reports a new analysis of the association between visual short-term memory (VSTM) and the brain-regional volumes of the hippocampi, primary visual cortices, and other cortical regions among vivid and non-vivid visual imagers. In a sample of 53 volunteers aged 54 to 80 with MRI scans, the performance of 10 Low VVIQ scorers and 10 High VVIQ scorers was compared. The subject groups included an aphantasic person and a hyperphantasic person. The High-VVIQ group yielded (i) significantly more accurate VSTM performance and (ii) significantly larger volumes of the hippocampi and primary visual cortex. For 47 subfields, the volumes in the hyperphantasic person exceeded those in the aphantasic person by an average of 57 percent. The study confirms the existence of robust empirical associations between VMI vividness, short-term memory, the regional volume of hippocampal subfields, and V1 area.

Coda

The aim of the Special Issue is to provide new perspectives on the role of visual mental imagery in how we image the world, both present and future. This network of international researchers has demonstrated a vital, dynamic, and flexible role for VMI in mental life. Three primary functions of VMI include (A) remembering recent- and distant-past *stimuli, episodes, events and scenarios* ("SEES"); (B) anticipating, foreseeing, and simulating near- and distant-future SEES; and (C) constructing phantasy for dreams and imaginary SEES. A key issue to investigate in future research is how, in rare cases (in this Guest Editor's opinion), voluntary VMI can be absent, and to what extent VMI absence could be effectively compensated for.

We look forward to further exciting discoveries that will help us to understand how we image the world.

Acknowledgments: The author wishes to thank the contributors and the reviewers, who facilitated the production of this Special Issue.

Conflicts of Interest: The authors declare no conflict of interest.

List of Contributions:

1. Marks, D.F. Phenomenological Studies of Visual Mental Imagery: A Review and Synthesis of Historical Datasets. *Vision* **2023**, *7*, 67. https://doi.org/10.3390/vision7040067.
2. Dijkstra, N. Uncovering the Role of the Early Visual Cortex in Visual Mental Imagery. *Vision* **2024**, *8*, 29. https://doi.org/10.3390/vision8020029.
3. Jin, F.; Hsu, S.-M.; Li, Y. A Systematic Review of Aphantasia: Concept, Measurement, Neural Basis, and Theory Development. *Vision* **2024**, *8*, 56. https://doi.org/10.3390/vision8030056.
4. Blomkvist, A. Shaping the Space: A Role for the Hippocampus in Mental Imagery Formation. *Vision* **2025**, *9*, 2. https://doi.org/10.3390/vision9010002.
5. Hatakeyama, T. Associations between Autistic-like Traits and Imagery Ability. *Vision* **2024**, *8*, 13. https://doi.org/10.3390/vision8010013.
6. Motoyama, H.; Hishitani, S. The Neural Basis of a Cognitive Function That Suppresses the Generation of Mental Imagery: Evidence from a Functional Magnetic Resonance Imaging Study. *Vision* **2024**, *8*, 18. https://doi.org/10.3390/vision8020018.
7. El Haj, M.; Moustafa, A.A. In the Eyes of the Future: Eye Movement during Near and Distant Future Thinking. *Vision*. **2024**, *8*, 32. https://doi.org/10.3390/vision8020032.
8. Pechenkova, E.; Rachinskaya, M.; Vasilenko, V.; Blazhenkova, O.; Mershina, E. Brain Functional Connectivity During First- and Third-Person Visual Imagery. *Vision* **2025**, *9*, 30. https://doi.org/10.3390/vision9020030.

9. Bilzer, M.; Monzel, M. The Phenomenology of Offline Perception: Multisensory Profiles of Voluntary Mental Imagery and Dream Imagery. *Vision* **2025**, *9*, 37. https://doi.org/10.3390/vision9020037.
10. Marks, D.F. Stronger Short-Term Memory, Larger Hippocampi and Area V1 in People with High VVIQ Scores. *Vision* **2025**, *9*, 53. https://doi.org/10.3390/vision9030053.

vision

MDPI

Review

Phenomenological Studies of Visual Mental Imagery: A Review and Synthesis of Historical Datasets

David F. Marks

Independent Researcher, 13200 Arles, Bouches-du-Rhône, Provence-Alpes-Côte d'Azur, France; dfmarksphd@gmail.com

Abstract: This article reviews historically significant phenomenological studies of visual mental imagery (VMI), starting with Fechner in 1860 and continuing to the present. This synthesis of diverse VMI phenomenological studies in healthy adults serves as a unique resource for investigators of individual differences, cognitive development and clinical and neurological conditions. The review focuses on two kinds of VMI, "memory imagery" and "eidetic imagery". Ten primary studies are drawn from three periods of the scholarly literature: early (1860–1929), middle (1930–1999) and recent (2000–2023). It is concluded that memory and eidetic imagery are two forms of constructive imagery, varying along a continuum of intensity or vividness. Vividness is a combination of clarity, colourfulness and liveliness, where clarity is defined by brightness and sharpness, colourfulness by image saturation and liveliness by vivacity, animation, feeling, solidity, projection and metamorphosis. The findings are integrated in a template that specifies, as a tree-like structure, the 16 properties of VMI vividness in healthy adult humans. The template takes into account the weight of evidence drawn from the accounts and reveals an extraordinary degree of consistency in reported VMI characteristics, revealed by specialized studies of healthy adult humans across time, space and culture.

Keywords: phenomenology; visual mental imagery; after-imagery; eidetic imagery; memory imagery; vividness; clarity; colourfulness; liveliness; projection; metamorphosis; individual differences

1. Introduction

1.1. Purpose

Any stable environment of familiar objects evokes sensations, images and feelings that vary by degree across individuals. Accordingly, it is surprising that any pair of individuals is able to reach agreement about what they are perceiving and feeling based on the two sets of fallible conscious experiences. While everyone may have their own unique perspective, there must also exist rules that create consistency. Anyone can experience the "redness" of an apple, the "pointedness" of a stick or "the warm feeling" of a campfire because there is, one must believe, something that it is like to have these experiences. There are particular phenomenological qualities or "qualia" that each of us has with these experiences [1]. The redness, roundness and warmth that we experience share sufficient commonality and consistency to enable us to communicate about them. What appears curious about this surprising fact rests not only with the mysterious qualia themselves, but with the interesting fact that qualia are triggered by sensorial objects and also by absent objects that are available only as mental images. Three tantalizing questions remain to be answered: what is it like to experience mental imagery, what is mental imagery for, and how are we to explain the striking individual differences? Books and journals are filled with accounts of theories and experiments, but a cohesive description of the phenomenology of mental imagery appears lacking. In attempting to answer the above-mentioned questions, one needs to review the findings of phenomenological studies, which is the purpose of the present review.

Almost every reviewer of mental imagery research starts with Aristotle (384–322 B.C.E.), who stated that "Images belong to the rational soul in the manner of perceptions, and whenever it affirms or denies that something is good or bad, it pursues or avoids. Consequently, the soul never thinks without an image" [2]. Skipping forward 22 centuries, we are not surprised to discover a few complications and technical hitches including, for example, Maurice Merleau-Ponty's assertion that "we can understand that nuances and sensation, perception and imagery are all ambiguous terms, even for what the psychologist of perception appears as an essence, e.g., figure vs. ground, can implode with the pressure of ambiguity" [3]. On occasion, philosophers appear confused about mental imagery, and one can discover perplexing claims such as the following: "It does not seem like mental imagery is an ordinary language term…and no languages other than English has (sic) a term that would mean mental imagery (as distinct from 'imagination' or 'mental picture')". The encyclopaedia continues, "This encyclopaedia entry will not attempt to give an ordinary language analysis of the term "mental imagery", partly because it is far from clear that 'mental imagery' is part of the ordinary language" [4]. These statements appear perplexing because, to the best of the author's knowledge, the leading world languages manifestly do all have a term for "mental imagery", e.g., Chinese: 心理意象 (929 M speakers), Spanish: "imágen mental" (475 M speakers), Hindi: मानसिक कल्पना (422 M speakers), Arabic: "الذهنية الصورة" (313 M speakers), French: "imagerie mentale" (300 M speakers), Russian: мысленные образы (258 M speakers) and German: "mentales Bild" (155 M speakers). Moreover, as is the case for English, these seven languages all use a different word for "imagination". Thus, for billions of speakers, the term "mental imagery"—or its precise lexical equivalent—is a part of the ordinary language, and the term "mental imagery" is distinct from the term for "imagination". I do not dwell any further here on philosophical views but move on to consider the foundations of our knowledge concerning the phenomenology of visual mental imagery (VMI).

The study of visual mental imagery since the so-called "Cognitive Revival" has received the theoretical attention of multiple psychologists [5–14]. However, without any agreement, clarity or precision about the phenomenological nature of visual mental imagery experiences, mental imagery investigators are acting like the blind monks examining the different parts of an elephant in the well-known parable (Figure 1).

Figure 1. The parable of the blind monks inspecting the different parts of an elephant (Public Domain).

My purpose here is to synthesize a collection of individual studies of the "elephant" that is visual mental imagery in an integrative review of VMI phenomenology. This review encompasses studies utilizing a variety of procedures for imagery evocation, including introspective analyses by competent observers, imagery volunteered in response to questionnaires, perceptual presentations, special instructions, interviews, or spontaneous evocations while listening to, or reading, texts. The majority of included studies are not experiments—although some were called such—but they have generated a sufficiently detailed description of imagery content to enable principles and processes to be compared and contrasted across participants and studies.

The ability to compare and contrast people's accounts of VMI is necessary if we are to progress our knowledge about the nature and purpose of VMI. Elsewhere, I have suggested that mental imagery is a foundation of conscious experience with six constituent modular processes: schemata, objects, actions, affect, goals and others' behaviour [15]. The evidence supporting these six connected modules and their interconnections provides empirical support from a wide range of studies, including neuroscientific studies, which show that all six modules of perception and mental imagery are associated with wide individual differences in the production of vividness [15]. The goal here is to explore the nature of produced VMI through phenomenological studies.

Traditionally, researchers have divided VMI into four kinds: after-images, eidetic images, memory images and imagination images, e.g., [16]. As noted, this review focuses on two of these four kinds, eidetic imagery and memory imagery, although necessarily there is some cross-over and "blending" between these two kinds and the other two kinds. As we shall see, the qualities, content and characteristics of VMI can be benchmarked against the objects of perceptual imagery.

Mental imagery research is replete with disciplinary shifts of interest, theoretical debates and bouts of "literature blindness", when the same phenomenon can be "discovered", re-labelled and republished over and over again. In the mental imagery field, one could say that there has been a surplus of theoretical assumptions and laboratory experiments but a dearth of phenomenological studies, which, one can only surmise, has been detrimental to progress. To the best of this author's knowledge, this is the first review of the psychological literature on VMI phenomenology covering the period from Fechner's 1860 landmark work to the present day. It is helpful to define the terms used here in reference to VMI.

1.2. Glossary of Terms

After-image (or perceptual after-image): one of four main kinds of mental image, in which an image is retained after the cessation of a stimulus that is externally projected, according to the principles of Emmert's Law.

Brightness (opp. dim): objects with the appearance of good illumination.

Clarity: a combination of brightness and sharpness, i.e., having a distinguishable outline and details.

Corporeity: the appearance of solidity, of having or being a body.

Easel Test: a procedure for testing the ability to produce eidetic images by presenting and then removing a picture on an easel for a fixed time.

Eidetic image: from the Greek "eidos", meaning idea or something seen, a persistent, vivid, projected image known to be subjective, so not confused with real objects.

Emmert's Law: as described by Emmert in 1881, retinal images that are perceived to differ in physical size when localized at different distances according to the formula: Perceived size = Retinal image size × Perceived distance.

Hallucination: a persistent, vivid, projected image that is mistakenly thought to be a real object.

Imagination image: a mental image of an object or combination of objects never previously experienced.

Individual differences: reliably observed differences between individuals in any mental, behavioural or physical attribute.

Introspective report: a description of mental experience within a particular set and setting.

Liveliness: a combination of vivacity, projection, solidity, animation, feeling and metamorphosis.

Memory image: a mental image of something previously experienced.

Mental image: a quasi-perceptual experience of an object or activity in the absence of the object or activity.

Metamorphosis: unplanned changes in image content that follow a sequence based on colour, form and meaning.

Multi-modal mental image: mental imagery that includes visual and/or auditory, olfactory, gustatory and tactile elements.

Open Circle Test: a procedure for testing a person's ability to produce eidetic images while looking at a circle, combined with hearing a stimulus in the form of a colour name.

Phenomenological study: an investigation of a person's experience of any form of mental activity.

Projection: the impression of an external spatial location and three-dimensionality.

Self-report: a person's description of a mental experience.

Sharpness (opp. vague): the quality of having a distinct outline and details.

Vivacity: the quality of being animated, affect-laden and active.

Vividness: a combination of clarity, colourfulness and liveliness.

1.3. Methods and Ground Rules

Methods for the phenomenological investigation of VMI have varied according to ideological positions about the nature of psychology that have held favour at different times. Typically, data have consisted of introspective reports by research participants in response to inquiries, interviews, questionnaires and "think-aloud" techniques in experimental settings (see Glossary, Section 1.2 above). More rarely, non-verbal methods such as drawing, dance, play, games, music and performance have been utilized. Here, some use is made of drawings. However, for practical reasons, VMI in the context of artistic media could not be included in this review. The review also necessarily avoids the intrinsic difficulties of mental imagery studies with children. Although research has often favoured the use of children, their questionable introspective abilities and the potential for the biasing of accounts with demand characteristics can be significantly reduced by employing adult participants. This review is concerned with the VMI of healthy adults, and research by Binet [17], Charcot [18] and Piaget [19] on child development, psychopathological and neurological conditions is not included.

The purpose is to construct a synthesis in the form of a "template" of VMI in healthy adult humans. Odd though it may seem, given the intensive efforts of psychologists to investigate mental imagery, our core knowledge about the content and characteristics of VMI has yet to be fully documented. Until now, no synthesis of knowledge about the range of content and characteristics of VMI in the healthy adult population has been produced. Without a template for the normal range of individual VMI capabilities, characteristics and contents, specialists in child development, psychopathology or neurological conditions are, in one real sense, working "in the dark" because they lack a normative standard.

The approach is to review "classic" and contemporary studies that have evaluated the visual mental imagery experiences of individuals to standardized introspective interviews, procedures and questionnaires. To be included in this review, an investigation must have recorded the participants' self-reported, i.e., verbal or non-verbal description, VMI content in response to a standard test, questionnaire, interview or other data-collection procedure. Random, anecdotal reflections do not meet the criteria. The datasets from the selected studies are quantitatively evaluated using objective criteria such as frequencies, averages and total scores for criteria that were defined by the original investigators themselves.

For example, Fechner ascertained his participants' answers to questions regarding whether it was easier to form visual mental images with their eyes open or closed, whether there was vividness or colourfulness to some degree, were there any feelings of pressure, contraction or effort, and whether it was possible to project the image to an external location. Reading Fechner's accounts, it is possible to categorize participants' responses according to each criterion, to tabulate the data and to arrive at an overall profile of the sample's distribution of responses. Galton focused his inquiry on illumination—is the image dim or fairly clear? Is its brightness comparable to that of the actual scene? Definition—are all the objects pretty well defined at the same time, or is the place of sharpest definition at any one moment more contracted than it is in a real scene? and colouration. Other criteria include the ability to project mental imagery into space, three-dimensionality, relative size, multi-sensorial attributes, movement, controllability, emotion or feeling and somatic aspects such as tension in or around the eyes.

A similar procedure must be applied to each and every dataset. Whenever possible, the same or similar criteria would be sought in different investigations, as would a distribution of participants' responses that could be tabulated, compared and aggregated across studies. Using these procedures, it is possible to harmonize and integrate different datasets into a criterion-based template in the form of a hierarchical description of visual mental imagery phenomenology, based on the investigators' and research participants' language about mental imagery experiences.

The studies to be reviewed are drawn from three historical periods: early (1860–1929), middle (1930–1999) and recent (2000–2023), which are considered in turn. Three of the leading investigators, together with Charles Darwin, a participant in one of the early studies, are depicted in Figure 2.

Figure 2. (**Top left**): Gustav Fechner, 1801–1887, the leading early investigator of VMI (Public domain); (**Top right**): Charles Darwin, 1809–1882, Galton's half-cousin, and a participant in Galton's studies included here (Public domain); (**Bottom left**): Peter McKellar, 1921–2003, a leading investigator from the middle period (photograph from the author's collection); (**Bottom right**): Takao Hatakeyama, Professor Emeritus of Yamagata University, Japan, a leading investigator from the middle period (photograph from the author's collection, © David F Marks).

2. The Early Period: 1860–1929

2.1. Gustav Fechner's (1860) Observations on After-Images and Memory Images

The phenomenological study of mental images began in earnest with the renowned German scientist, Gustav Fechner [20]. As the founder of psychophysics and co-developer of the Weber–Fechner Law, Fechner presented the world with a unique treasure trove of phenomenological observations on VMI that has been sorely neglected by anglophones. In his two-volume work, *Elemente der Psychophysik* (1860), Fechner stated the following:

> *"Initially, and in general, it cannot be denied that the mental sphere is subject to quantitative considerations. After all we can speak of a greater or lesser intensity of sensation; there are drives of different strengths, and greater and lesser degrees of attention, of the vividness of images of memory and fantasy, and of clearness of consciousness in general, as well as of the intensity of separate thoughts".* ([20], p. 46).

Until now, only Volume I of Gustav Fechner's (1860) *Elemente der Psychophysik* had been available in English [20]. Here, I review Fechner's account of VMI in Volume 2 of *Elemente der Psychophysik*, chapter XLIV [21]. Fechner writes:

> *"If the memory-images, fantasy-images and schemata accompanying thought are all still psychophysically founded, so is thought itself, in that every other substance and course of thought presupposes another material and another way of linking the schemata, without which no thought can take place at all, just as another melody and harmony cannot be without other tones and another way of linking the tones. Now a piano, with its comparatively small number of fixed keys, nevertheless affords the possibility of executing the most diverse melodies and harmonies, and however many and however high thoughts man may conceive, 25 letters suffice to express them; in both cases it depends only on the connection and the sequence in which the keys or letters are passed through. The brain, however, with its innumerable fibres, active in various ways, contains incomparably richer means in this respect, so there can be no obstacle to trusting it with at least as great a performance inwardly as we perform outwardly by means of it".* [20].

Fechner's observations [21] concerning after-imagery, memory imagery and comparisons between them, were based on interviews with six professorial and professional colleagues, including his wife, and introspections of his own mental imagery. The recent translation of Fechner's [21] mental imagery research is prefixed with Fechner's remarks, as follows:

> *Sensory impressions once made from the outside continue to exist for a certain time after the removal of the external stimulus as after-images, after-sounds, generally as after-perceptions, which in a healthy, strong state of the senses tend to be less easily perceived, less intense and lasting than in a weakly stimulable state; and they leave behind the capacity to be reproduced in memories or more or less transformed in phantasy images. Both kinds of after-effects are to be considered here mainly, if not exclusively, in the field of facial perception, where they have been most studied; but what is valid here is more or less applicable to other fields of sense perception.*
>
> *The main differences between after-images on the one hand, and memory and phantasy images on the other, are that the former are only ever accompanied by a feeling of receptivity. The first always arise and exist only with a feeling of receptivity, only in continuity with the sensory impressions made, independent of volition and association with the imagination, and, according to the immediately preceding sensory impressions, also proceed independently of volition, legitimately, whereas the memory and phantasy images, with a feeling of lesser or greater spontaneity, can arise even a long time after preceding sensory influences, partly involuntarily through association with the imagination, partly volitionally, and can be banished and altered again.* ([21], pp. 3–4).

Thus, Fechner makes a distinction between the after-image as *involuntary*, and memory and imagination images as *volitional* or *partly volitional*, and so can be "extinguished" by an act of will or the needs of the moment. Fechner discusses his own imagery ability as follows:

> *In general, memory and fantasy images always appear to me as something lacking in corporeality, airy, breathy, in contrast to the more material impression of the after-images.*
>
> *Thus, the drawing of the memory and fantasy images is more vague and blurred than that of the after-images. I am not able to obtain clear, sharp outlines even on the most familiar memory images of the objects that are daily before my eyes, while the after-images appear with corresponding sharpness as directly seen objects.*

After-images in the closed eye are either deeper black or lighter than the surrounding ground of the eye and the uniform black of the field of vision, depending on the brightness of the objects viewed in relation to the ground on which they appeared. Memory images, on the other hand, generally give me a weaker impression than the black itself. From white to black there is a scale of continuously graduated brightness and the deepest black is the pure black of the eye. If I now ask myself where this scale would lead if I were to think of it as continuing below black, I believe that one is led to the indistinct impression of memory and fantasy images.

With all my efforts, I cannot reproduce colours in the memory images of coloured objects, or only in fleeting, doubtful appearances when recalling very striking impressions (for example, when I think of cut eggs on spinach, where the white, yellow and green stand out very sharply against each other) while I receive vivid coloured after-images in the open as well as the closed eye. I also never dream in colours, but all my experiences in dreams seem to me to proceed in a kind of twilight or night.

I am not able to recall even the most familiar memory-images.

It is not possible to hold on to the image steadily, even for a short time, but in order to look at it longer, it must, so to speak, always be recreated anew; it does not both change of its own accord and disappear again and again of its own accord. If, however, I want to reproduce it often one after the other with the same intention, it soon no longer succeeds at all, for the attention or activity of production soon dulls. This, however, is not a dulling of memory activity in general; for I am not prevented—and this seems to me worthy of attention—from immediately imagining another familiar memory image instead, as clearly as it is at all possible for me to do so, and, when attention or production activity has also exhausted itself for this one, to return to the first image where I can produce it again with the initial clarity. This is true even of quite related pictures; as, for example, I have often attempted with two portrait figures on the same photograph or portraits hanging next to each other in my living room, neither of which I can often reproduce in memory one after the other, but both in repeated alternation. If, however, I continue this alternation somewhat quickly and often one after the other, I finally find myself dulled for both pictures, but can pass on to a third picture with success.

I cannot change after-images at all by will. ([21], pp. 4–5).

For economy of space, I tabulate a summary of the remainder of Fechner's observations in Table 1.

Table 1. Showing the qualities of memory imagery reported by Fechner and six interviewees [21].

Identity	Age	Rank *	Easier with Eyes Open or Closed **	Vividness to Some Degree	Colour to Some Degree	Feelings of Pressure or Contraction or Effort ***	Project to An External Location	3 D	Multi-Sensory
G Fechner	59	=6	Open	Dim	Fleeting	+	+	+	+
H Weisse	59	=6	Open	Faint	Little	?	+	?	?
A W Volkmann	59	5	=	Faint	Faint	+	+	?	?
W Hankel	46	4	Open	Yes	+	+	+	?	?
M W Robisch	50s	3	=	High	+	+	+	+	+
C M Fechner	51	2	Closed	High	+	+	+	+	+
M Busch	39	1	?	Very high	+	+	+	+	+
Total	-	-	-	7/7	7/7	6/6	7/7	4/4	4/4

* Fechner's ranking of each observer's imagery strength. ** The participant's own judgment about whether their visual imagery was stronger with eyes open or closed or of equal (=) strength.*** Feelings of pressure or tension in or around the eyes and/or head. + indicates the participant reported the characteristic in the column heading. ? indicates the datum was missing.

From the results of his "experiments", Fechner noted the following:

"(1) Sometime after the creation, the figures disappear or change into others without my being able to prevent this. (Fechner's report of figures disappearing and changing into others without control matches the phenomena of metamorphosis described later in this paper).

(2) If the colour does not belong integrally to an object, I do not always have it completely under my control. A face, for example, never appears blue to me, but always in its natural colour, whereas instead of the imaginary red cloth, a blue one can appear at times; in general, the production of a certain colour is more difficult than that of a certain shape, and the first one succeeded in my p. 485

I did not succeed in the first attempts, since I had already succeeded in the last.

(3) I have succeeded a few times in seeing pure colours without objects; they then filled the entire field of vision.

(4) Objects that are not familiar to me, i.e., mere fantasy images, I often do not see, and instead of them, familiar objects of the same kind appear to me; for example, I once wanted to see a brass sword handle with a brass basket, but instead I saw the more familiar image of a rapping basket.

(5) Most of these subjective appearances, especially if they were bright, leave after-images if the eyes are opened quickly during the dwelling of the appearance; for example, I thought of a silver stirrup, and after looking at it for a while, I opened my eyes and saw the dark after-image of it for a long time." (Fechner's report of the after-imagery of memory imagery has been verified by later observers, e.g., [22,23]) [21].

In addition to seven cases, Fechner discusses accounts of mental imagery by figures such as Johann Wolfgang von Goethe, Gerolamo Cardano and others. These anecdotal observations presage findings reported by later investigators. For example, the first-person account attributed to Goethe, is golden:

"Goethe says in Contributions to Morphology and Natural Science: "I have the gift, when I close my eyes and, with my head bowed down, think of a flower in the center of my visual organ, it does not remain for a moment in its first form, but it spreads out and from within it unfolds again new flowers of coloured, even green, leaves; they are not natural flowers, but fantastic, but regular, like the rosettes of the sculptors. It is impossible to foresee the sprouting creation, but it lasts as long as I like, does not tire and does not intensify. I can produce the same if I think of the ornament of a colourfully painted disk, which then also changes continuously from the center to the periphery, completely like the kaleidoscopes." ([21], p. 18).

Goethe's account of the spontaneous metamorphosis of VMI content is the first known description of the phenomenon studied by psychologists two centuries later (see Section 3).

Fechner next discusses the account of Gerolamo Cardano (1501–1576), the Italian polymath and mathematician, "who tells of himself that he had been able to imagine luminously what he wanted". Then, Alexandre Brierre de Boismont (1797–1881), who discussed the mental imagery reported by an unidentified, prodigious portrait painter:

"One painter, who had inherited a large part of the clientele of the famous Sir Joshua Reynolds, and believed him to be of superior talent to his own, was so busy that he confessed to me, says Wigan, that he had painted 300 portraits, large and small, in one year. This fact seems physically impossible; but the secret of his speed and astonishing success was this: he needed only one sitting to represent the model. I saw him execute before my eyes in less than eight hours the miniature portrait of a gentleman whom I knew very well; it was done with the greatest care and a perfect likeness.

I asked him to give me some details of his process, and this is what he replied: "When a model came along, I looked at it attentively for half an hour, sketching on the canvas from time to time. I didn't need any longer. I would remove the canvas and move on to another person. When I wanted to continue the first portrait, I took the man in my mind, I put him on the chair, where I saw him as clearly as if he were '616' written in the sky; and I can even add with sharper and more vivid shapes and colours. I looked from time to time at the imaginary figure, and I began to paint; I suspended my work to examine the pose, absolutely as if the original had been before me; every time I cast my eyes on the chair, I saw the man." ([21], pp. 18–19]).

Fechner's findings warrant the following conclusions about VMI:

(i) Wide individual differences were evident in both vividness and colouration;
(ii) Some people reported stronger imagery with their eyes open, others with their eyes closed and others reported imagery of equal strength with their eyes open or closed;
(iii) Fechner's participants reported an impression of pressure, contraction or tension in generating VMI;
(iv) In all seven cases, VMI was projected into external space, corresponding to normal vision;
(v) In four cases, including Fechner himself, solid-looking, three-dimensional images could be formed;
(vi) In the same four cases, multi-sensory imagery spontaneously arose;
(vii) The cases Fechner presented comprised seven intellectuals, six men and one woman. As Fechner was aware, larger, more diverse samples would be required to form more definitive conclusions. The VMI characteristics in this study were evoked under Fechner's particular instructions and may not generalize to other procedures for inducing VMI, e.g., spontaneous VMI while reading, listening or thinking.

2.2. Galton's Breakfast Table Questionnaire

In 1880, a widely cited but misinterpreted study by Sir Francis Galton [24,25] reported findings obtained with his "Breakfast Table Questionnaire", which Galton gave to 100 "men of science", including his half-cousin, Charles Darwin, and numerous friends and colleagues. Galton also administered the same questionnaire to 172 schoolboys, whom he divided into two groups, "to serve as a check upon one another". Group A comprised boys of the four upper classes in the school, group B those of the five lower classes. The first questions were presented as follows:

"Before addressing yourself to any of the Questions on the opposite page, think of some definite object—suppose it is your breakfast-table as you sat down to it this morning—and consider carefully the picture that rises before your mind's eye. (p. 302).

1. Illumination.—Is the image dim or fairly clear? Is its brightness comparable to that of the actual scene?

2. Definition.—Are all the objects pretty well defined at the same time, or is the place of sharpest definition at any one moment more contracted than it is in a real scene?

3. Colouring.—Are the colours of the china, of the toast, bread-crust, mustard, meat, parsley, or whatever may have been on the table, quite distinct and natural?"

Wide individual differences in the vividness, projection and corporeity of responses were obtained, which Galton divided into three categories: cases where the faculty is *very high*, mediocre or *at the lowest*. One curious feature of Galton's own statements about his findings that has received critical comment was his claim to have been "amazed" at the poor or non-existent ability of the sample of scientists to mentally image. In fact, the results show the exact opposite. In only less than a handful of cases—perhaps only one out of one hundred—was zero imagery power reported. The vast majority of 99% reported at least some imagery ability. We return to this point later.

The first three of the "very high" category stated the following:

1. Brilliant, distinct, never blotchy.

2. Quite comparable to the real object. I feel as though I was dazzled, e.g., when recalling the sun to my mental vision.

3. In some instances quite as bright as an actual scene.

In the "mediocre" category, the first three people stated the following:

46. Fairly clear and not incomparable in illumination with that of the real scene, especially when I first catch it. Apt to become fainter when more particularly attended to.

47. Fairly clear, not quite comparable to that of the actual scene. Some objects are more sharply defined than others, the more familiar objects coming more distinctly in my mind.

48. Fairly clear as a general image; details rather misty.

The last three cases in the lowest category stated the following:

98. No. My memory is not of the nature of a spontaneous vision, though I remember well where a word occurs in a page, how furniture looks in a room. The ideas are not felt to be mental pictures, but rather the symbols of facts.

99. Extremely dim. The impressions are in all respects so dim, vague and transient, that I doubt whether they can reasonably be called images. They are incomparably less than those of dreams.

100. My powers are zero. To my consciousness there is almost no association of memory with objective visual impressions. I recollect the breakfast table, but do not see it.

To compare different samples along comparable scales, Galton [24] invented a system whereby each of his three samples were divided into octiles. A high degree of consistency was obtained across octiles from the three samples. Galton's findings are summarized in Table 2.

Table 2. Descriptive information on voluntary visual imagery by Galton's three British samples, together with Armstrong's US replication.

Sample	Complete or Partial Vividness	Colour Relatively Distinct and Natural	Extent of Field of View Larger Than, or Same as, Normal	External Projection Corresponded to Reality, or in Front of Eyes, at Least on Some Occasions
Galton's 100 men of science	97/100	97/100	75/100	Missing data
Galton's sample of schoolboys	167/172 ^	167/172 ^	47/121 *	126/160 **
Sub-totals	264/272 97.0%	264/272 97.0%	122/221 55.2%	126/160 78.8%
Armstrong's US students	183/188 97.3%	183/188 97.3%	95/188 50.5%	153/188 81.4%
Totals	447/460 97.2%	447/460 97.2%	217/409 53.0%	279/348 80.2%

^ Estimate. * In 51 missing cases, the description was "insufficient". ** Twelve missing data.

Galton's observations permit the following conclusions:

(i) Similar to Fechner's data, Galton's samples reported wide individual differences in the vividness and colourfulness of their VMI;
(ii) Also in line with Fechner's data, some people reported stronger imagery with their eyes open, others with their eyes closed;
(iii) Consistent with Fechner's findings, in a large majority of cases, mental images were localized in external space, corresponding to normal vision;
(iv) In the majority of cases, the apparent field of view in mental imagery was enlarged or the same as normal vision;

(v) In several cases, three-dimensional visual imagery could be formed, again confirming Fechner's findings;
(vi) In at least some cases, multi-sensory images were formed.

Galton's observations were limited to men of high socio-economic status, so they lacked diversity, limiting the generality of the conclusions to the human population as a whole. As noted above, Galton mis-stated his findings in alleging a lack of mental imagery ability in his sample of male scientists [24,25]. Galton's "amazement" was plainly inconsistent with the facts. Galton's observations did not in the very least support his claims about the deficient visual imagery of scientists. In a literature review, Brewer and Schommer-Atkins [26] found close to 50 secondary sources from 1883 to the present day that repeated, in parrot fashion, Galton's conclusion that scientists were totally lacking in visual imagery or had only "feeble" powers of mental imagery. This episode indicates a curious tendency in academic psychology towards "literature blindness", a kind of cognitive myopia, in which past errors are repeated without correction. Galton's conclusion is contradicted by the responses of Charles Darwin, who described the definition of his VMI "as distinct as if I had photos before me" and "perfectly coloured" [26].

Brewer and Schommer-Atkins attempted to replicate Galton's study with contemporary scientists and undergraduates and found *none* totally lacking in visual imagery, and only very few with feeble visual imagery. Brewer and Schommer-Atkins stated that Galton's conclusions were a "theory-laden interpretation of data based on the initial responses from several very salient scientists who reported little or no visual imagery on Galton's imagery questionnaire" ([26], p.130).

2.3. *Armstrong's Replication of Galton's Studies*

In the early 1890s, A. C. Armstrong [27] replicated Galton's Breakfast Table study in the USA, with the assistance of C H Judd. The investigators gave Galton's original questionnaire to 188 men, 37 at Princeton College (1881–1882) and 151 at Wesleyan University (1890–1893) [24]. According to their answers about "illumination", Armstrong divided the sample into five classes: "This class-division based on 1. is in general one of the most important that can be made" ([27], p. 498). The answers to Galton's question four—"What difference do you perceive between a very vivid mental picture called up in the dark and a real scene?"—showed "a close connection between the degree of illumination and the sense of reality". Retaining the same five-fold division for question two, Armstrong found two "natural" subdivisions under each class: "A, those subjects who have all the objects on the imaged table pretty well defined at the same time, and B, those whose place of sharpest definition is somewhat contracted". The replies to question three indicated, again, a widespread individual difference in the visualization of colour, with 172 of 188 subjects (91.5%) thinking that their colour imagery was relatively "distinct and natural", while only 16 (8.5%) failed in this respect although a few of these could "bring out the missing quality by turning their attention toward it".

The answers to question five also showed remarkable consistency with those obtained by Galton: "Distance of Images.—Where do mental images appear to be situated—within the head, within the eyeball, just in front of the eyes, or at a distance corresponding to reality? Can you project an image upon a piece of paper?". The majority of Armstrong's students (133 or 71%) stated that the distance corresponded to reality, while six saw their images at positions more distant than the real scene, "as in a kind of bird's eye view". Of the remaining 49, 19 localized their images in the head, 14 just in front of the eyes, 4 in the eyeballs, and 12 at variable distances.

As part of his study, Armstrong included a "Mr. A. G. C", a gifted visualizer, who was not in the student sample. Armstrong states the following: "His images are so habitually localized as in reality" and "he is also endowed with the gift of mentally 'seeing around objects', and he could image four faces of a die without difficulty. The visualization of all six faces, however, costs him an effort of attention, and when he was asked to accomplish it, it was noted that the hand which he held in front of him to aid his visual imagination (as though the die were held in it) *moved up toward the eyes*. Questioned about the fact, Mr. C. recognized the movement, although he had performed it involuntarily, and was able to estimate it in inches since his business and habits accustom him to measurement. Without strain he holds the hand and visualizes images at about 14 inches from the eyes; with strain they move up to a distance of 4 inches".

Commenting on his study, Armstrong noted that "the most striking phenomenon shown is the intimate relation of imagery and attention and the effect of the latter on the various phases and characteristics of the former". He noted that women students from Wesleyan and Vassar College showed a "higher development of the faculty" of mental imagery than the sample of men ([27], p. 506). Armstrong's findings from his sample of men fully replicated those of Galton and double the weight of the latter's conclusions.

2.4. Fernald's Studies

Mabel Ruth Fernald presented her PhD dissertation in 1912 at Princeton University, USA, on "The Diagnosis of Mental Imagery" [28]. With references to Fechner (1860), Galton (1880, 1883) and Armstrong (1894), Fernald's dissertation contained no less than 118 references, which is indicative of the intense scholarly interest in mental imagery during this period. For her own introspective study, Fernald chose 11 participants who were instructors or advanced graduate students trained in introspection. The nature of that training was not specified. Five were men (A, Ad, Hs, P and S) and six were women (C, G, H, Sun, T and V). For the first time in a phenomenological study, the investigator and majority of participants were women. The tests were conducted at the University of Chicago during the period of 1908 to 1910.

The participants were given a variety of "diagnostic" tests, and the current review is confined to the results for reading. In contrast to Fechner, Galton and Armstrong, who studied voluntary imagery evoked by instructions or questionnaires, Fernald's test procedures evoked imagery that occurred spontaneously while reading evocative passages. Twenty-one short passages were used with descriptive material in varying levels of abstractness–concreteness and referring to varied sensory fields. Fernald's procedures included silent reading at a normal rate with the content reproduced, silent reading at faster rates with the content reproduced, silent reading at a normal rate without reproduction, the passage read out loud by the investigator and the passage read aloud by the participant. The participants reported their introspections on (a) the verbal imagery used; (b) the illustrative imagery most commonly used; and (c) a statement on the nature and amounts of these two imagery kinds. Table 3 shows the findings for each participant, including their overall imagery ability as rated by Fernald, and the incidence of imagery of six different kinds for each participant over the readings. The range of variation is moderately high, with the incidence of tactile, organic or visceral imagery reported by all 11 participants, and temperature (warmth, cold) by five. It is notable that eight of the eleven participants reported at least five of the six kinds of imagery, with one participant (H) rated as having "minimal" imagery ability, reporting only three of the six kinds.

Fernald's findings enable the following conclusions:

(i) The range of individual differences was large. However, every one of the eleven participants reported spontaneous illustrative mental imagery while reading;
(ii) Even participants who reported scant or minimal amounts of illustrative imagery still experienced reading as illustrated by mental imagery of at least three kinds;
(iii) The high incidence of tactile, organic or visceral imagery, which was reported by all participants, was striking;
(iv) Visuo-motor imagery was prevalent, with only one of the eleven participants reporting an absolute lack of visuo-motor illustrative imagery;
(v) In line with (iv), coloured imagery was present for ten of the eleven participants;
(vi) Auditory imagery was present in all but three cases (H, S and Sun) but, overall, less frequently than visual imagery;

(vii) The sample size was limited, and potential gender differences could not be statistically tested with the data from this study.

Table 3. Illustrative imagery while reading as reported by Fernald's participants [28].

ID	Gender	Individual's Overall Imagery Ability, as Rated by the Investigator	Tactile, Organic or Visceral Imagery	Visuo-Motor Imagery	Colour	Olfactory	Auditory Imagery	Temperature	Totals
A	Man	Minimum of illustrative imagery	+	+	+	+	+	No	5/6
Ad	Man	Minimal	+	+	+	+	+	+	6/6
C	Woman	Profuse visual imagery. Other imagery moderate.	+	+	+	+	+	No	5/6
G	Woman	Moderate amount of illustrative imagery	+	+	+	+	+	No	5/6
Hs	Man	Abundant imagery	+	+	+	+	+	No	5/6
H	Woman	Minimal	+	No	+	+	No	No	3/6
P	Man	Fair amount, but vague and indefinite	+	+	+	+	+	No	5/6
S	Man	Profuse, vivid and detailed	+	+	+	No	No	+	4/6
Sun	Woman	Scanty	+	+	No	+	No	+	4/6
T	Woman	Scanty, lack of detail, not vivid	+	+	+	No—only sniffed	+	+	5/6
V	Woman	Profusion of non-visual imagery	+	+	+	+	+	+	6/6
TOTALS			11/11	10/11	10/11	9/11	8/11	5/11	53/66

+: Indicates a positive report of a characteristic by a participant.

As a general comment, we should not be surprised that Fernald pointed to a difficulty of articulation for "even the best" of her introspectionists, whom "had been so absorbed in the operation that they could not tell how they had done it" ([28], p. 136). This question is raised repeatedly throughout phenomenological research and is impossible to extinguish because it is palpably valid. The method of introspection fell into disuse after 1913, when the American behaviourist John B. Watson proposed that psychology could manage without it and any other method for research on consciousness [29]. Watson advocated a psychology that he claimed was "a purely objective branch of natural science", and behaviourist methods with animal experiments on learning and memory replaced introspection. As a consequence, almost no phenomenological studies occurred in psychology between 1930 and 1957, when a book by TPH (Peter) McKellar called *Imagination and Thinking* appeared [30]. We turn next to VMI studies concerned with eidetic imagery.

2.5. Early Studies of Eidetic Imagery

A starting point for studies of eidetic imagery is Erich R. Jaensch [31,32], working at Marburg, Germany, who defined eidetic imagery in 1923 thus:

> *"Optical perceptual (or eidetic) images are phenomena that take up an intermediate position between sensations and images. Like ordinary physiological after-images, they are always seen in the literal sense. They have this property of necessity and under all conditions, and share it with sensations. In other respects they can also exhibit the properties of images (Vorstellungen). In those cases in which the imagination has little influence, they are merely modified afterimages, deviating from the norm in a definite way, and when that influence is nearly, or completely zero, we can look upon them as slightly intensified after-images. In the other limiting case, when the influence of the imagination is*

> *at its maximum, they are ideas that, like after-images, are projected outward and literally seen."* ([31], pp. 1–2).

Jaensch's statement continues:

> *"For the great majority of adults there is an unbridgeable gulf between sensations and images. It has always been known that for a few individuals this is not true. Some people have peculiar ' intermediate experiences ' between sensations and images. From the description that such people have given of these experiences, and from the characterization we have just given of eidetic images, we must conclude that their ' experiences ' are due to eidetic images. These phenomena, it is true, are rare among average adults."* ([31], pp. 3–4).

According to Jaensch [31], "a positive EI is a strong form of visual after-image having great clarity and in the colours of the stimulus object, often in very fine detail". In Jaensch's view, "the frequency of eidetic imagery is highest in young children. Its opposite, the negative EI, appears in complementary colours".

Jaensch used the observed individual differences in eidetic imagery ability to develop a speculative typological system, which Jaensch opportunistically altered and aligned with the racist philosophy of national socialism in 1930s Germany. Attempts by independent psychologists to replicate Jaensch's findings failed, and the resulting theoretical controversy led to Jaensch's work being becoming discredited, e.g., [33,34].

Allport (1924) [33] pointed out that eidetic images are definitely not photographic because of their "flexibility", in which they frequently metamorphose with extensive, unconscious and uncontrolled changes, and have a frequent incompleteness of contents. It had also been reported that eidetic imagery may arise spontaneously with no specific evoking perceptual stimulus [31,32]. The Marburg School used the images' spontaneous appearance as an indicator of the "B-type". In the case of the "B-type", according to Klüver [34], "the eidetic image is often nothing but a visualized idea projected into perceived space... He can, without any effort, produce eidetic images and, at pleasure, vanish them; he can do this without a preceding presentation of a stimulus..." (p. 181). In the next section, eidetic imagery phenomenology is discussed in more depth and detail.

3. The Middle Period: 1930–1999

The middle period began in what—from the perspective of phenomenological research—could be called "The Dark Ages". For 50 years, from 1913 to 1963, phenomenological research on human mental life was effectively "cancelled" and replaced by experimental, behavioural studies of learning and memory. For mainstream psychological science, mental imagery and consciousness were "beyond the pale" and were deemed not to exist. Ultimately, mental imagery research was actively resumed in the 1960s, when detailed phenomenological and experimental studies together began to reveal a high degree of consistency across studies. This section begins with two contrasting approaches to eidetic imagery, one viewing it as "reproductive" and the other viewing it as "constructive". The former approach gradually diminished towards a common approach based on the principle that mental images are phenomenological constructions.

3.1. The Myth of Eidetic Imagery as "Photographic" Memory

One hypothesis that originated with Jaensch and received continued attention was the idea that eidetic imagery is a so-called "photographic" memory, in which any visual stimulus—allegedly—can be held as a lasting, accurate, projected image. For example, Gengerelli [35] observed a young woman who could correctly discriminate tiny differences in the diameters of mentally imaged circles compared to a standard stimulus presented at some distance. The majority of studies on eidetic imagery have used the Easel Test. A blank piece of grey card, 24 by 30 inches, is placed on a support. First, the participant is approximately 20 inches in front of the screen, fixating on a red, four-by-four-inch square for 10 s, then, after removal of the stimulus, reports what they subsequently "see" on the screen. Successively, the procedure is repeated with blue, black and yellow square stimuli. Being used to seeing images—most people report negative after-images—the participant is shown a silhouette drawing, purely black-and-white, which they scan for one half-minute then, when it is removed, they indicate what they image on the screen. If they report an eidetic imagery, they are questioned about its content; otherwise, they are asked about the picture which they had just previously viewed. Three more pictures are presented in a sequence: one more silhouette and two full-colour pictures. The four pictures originally were chosen for North American children: silhouettes showing a family scene and an Indigenous American hunting (both containing animals); the coloured ones feature another Indigenous American fishing and a scene from "Alice in Wonderland" showing Alice with the Cheshire cat.

Based on studies with children of varying ages, the hypothesis of "photographic" memory has been investigated with generally inconsistent findings. Ralph N. Haber [36,37] used the Easel Test with five behavioural criteria for differentiating eidetic imagery from afterimagery. Eidetic imagery occurred when (i) saccades occur during the VMI of the stimulus, with the image remaining still; (ii) the image has a protracted duration, and appears in the original colours; (iii) the eidetic image is projected into space to fall on any available surface; (iv) the present tense is used by the participant when reporting images, not the past tense; (v) two images can be superimposed and reported as a composite [37]. During a five-year period, however, Haber struggled to find only 20 children (less than one per cent of his sample) with eidetic imagery as the so-called "photographic" type of memory, which forced the conclusion that *"eidetic images are only available to a small percentage of children 6–12 years old, and are virtually nonexistent in adults. However, extensive research has failed to demonstrate consistent correlates between the presence of eidetic imagery and any cognitive, intellectual, neurological, or emotional measure."* [37].

Haber devised a more direct test by showing two pictures in sequence, which, when superimposed, formed a third image showing a person's face. Four participants reported seeing the facial eidetic imagery, but this test was insufficiently stringent because the children might have guessed the composite face from either picture alone. Then, a Harvard student claimed she could "hallucinate at will a beard on a clean-shaven man or repeat a page of poetry written in a foreign language verbatim—backwards and forwards—after simply looking at it" [38]. Stromeyer and Psotka [39] attempted to demonstrate this person's ability using a new technique involving random dot stereograms. However, demand characteristics may have fallen into play and one investigator reported a photographic eidetic imagery memory incidence of "none in a million" [40] in the general population. The possible reasons for this failure will be indicated in the following sections.

3.2. *Cross-Cultural Study with Participants from Five African Countries*

Leonard W. Doob (1966) [41] conducted cross-cultural studies of eidetic imagery in five African countries, using Haber's methods with the Easel Test and culturally adapted materials. Doob found it difficult to consistently evoke after-images, and so he extended the exposure time to 15 s. Because Haber's pictures were "culturally saturated", Doob substituted photographs taken in local African settings such as a picture of African women with infants and metal rooves (sic) on their heads and a scene in colour of a bus with a full load of passengers. Additionally, complicated montages were used to test the accuracy of the detail of eidetic imagery or picture memory, e.g., a white elephant beside a white tree, with a multi-coloured car above, and a yellow giraffe and green elephant beside a yellow-coloured tree below. The examiner sat opposite the participant at a small table so they could see how well the instructions to keep the eyes fixed on the colours and to scan the pictures were obeyed, and whether the participant's attention was directed to the screen during the reporting of eidetic imagery. Virtually all testing occurred out-of-doors in varying, poorly controlled conditions. The investigator notes that it was field research requiring a standardization of procedures that would often have fallen short of achieving the methodological ideal.

Samples of adults were tested in five African societies: a group of Ibo in Eastern Nigeria (N = 28; 39%); a sample of Kamba in Central Kenya (N = 49; 39%); a nomadic group of Somali (N = 24; 4%); a nomadic group of Masai in Kenya (N = 20; 80%); and a mixed Swahili-speaking group in eastern Tanzania (N = 33; 15%). The percentages show the proportion of each sample reporting some eidetic ability. The following quotations represent four of the positive responders' experiences:

> *IBO CULTIVATOR: I see human beings. (How many are there, can you count them?) Yes, four [five].*
>
> *KAMBA WOMAN: I see some people and motor cars, one man on top, with goods on top. (What is the color of the bus?) Looks like red and green [brownish and white]. (What else do you see?) Many people.(Do you see any trees?) Yes. (Where?) [Points correctly]. (Do you see anything else?) I see nothing more. (Can you see the license plates on the cars?) No.*
>
> *MASAI HERDER: (Can you see their trousers?) No, I can't see them. (What do you see on the screen now?) I still can see that person, but it is not now very clear.*
>
> *SWAHILI CULTIVATOR: I see some numbers, an elephant, a man. (Do you see them clearly?) I see them. (Can you see the numbers?) I see the numbers clearly. (Can you read them?) I can read them. (Would you read them, please? Tell us what numbers you see on the screen now.) 0 [pause of 7 s] 2 [5 s] 2 [4 s] 1 [5 s] 7 [6 s] 0 again [actually: 0714653282760].* ([41], pp. 22–23).

Doob reported: "Almost without exception those with eidetic imagery...stated that they had been amazed to see the images on the screen, that this was an experience they had never or seldom previously had (Masai 7/9; Swahili 7/8; Kamba 22/25)". However, this latter finding was most likely a consequence of the participants' unfamiliarity with being tested by looking at pictures on easels because the majority claimed they customarily were used to "seeing pictures in front of my eyes" (projected imagery) (Somali 2/6; Masai 15/21; Swahili 18/32; Kamba children 45/57). There was no statistically significant relationship between the presence of eidetic imagery and of projected imagery in any of the groups. Likewise, many informants stated that often they have "pictures in my head," or pictorial images when recalling events or people from the past (Somali 6/9; Masai 13/14; Swahili 6/16; Kamba 66/86). Pictorial images were occasionally reported by Somalis to be "in my heart" and by the Swahili to be "in my mind". No association existed between eidetic imagery and pictorial imagery, nor between eidetic imagery and projected imagery. During or after the tests for eidetic imagery, the accuracy of recall was also found to be unrelated to claims of projected imagery or pictorial imagery for the pictures that had been exposed. A Somali stated that he could see "the bus" in his head but, when asked to describe what he saw on top of the bus, he incorrectly replied that "nothing" was there. From the standpoint of most informants questioned, there was a difference between remembering with eidetic, projected or pictorial imagery on the one hand, and what is ordinarily called "simple memory" or recall on the other (Somali 15/15; Masai 4/4; Swahili 7/8; Kamba adults 14/14). Doob's [41] observations lead to the following conclusions:

(i) The incidence of eidetic imagery is not significantly higher or lower in African societies than in European or US samples;
(ii) A high degree of variability was evident within each sample and "the attributes and characteristics of eidetic imagery when they do appear do not vary from society to society and, in fact, are no different from those found in Europe and America" ([41], p. 29);
(iii) As was the case in the Western laboratory studies, the differences within groups appeared much greater than those between groups.

3.3. A Rekindling of Phenomenological Research on VMI

With the exception of the above-mentioned studies on eidetic imagery, the absence of any new phenomenological studies for almost half a century left a vacuum that remained unfilled until the 1960s and 1970s, when an "explosion" of mental imagery studies occurred. The 1957 book by Peter McKellar, *Imagination and Thinking* [40], presaged a re-awakening of psychological interest in mental imagery. Between 1957 and 1975, a productive group of researchers made substantial contributions to our understanding of individual differences in VMI. In 1974, a Japanese psychologist Takao Hatakeyama [42,43] published a ground-breaking study of eidetic imagery in a 20-year-old woman (Y.K.). This fascinating case appeared as a pinnacle of eidetic imagery at the high ability end of the distribution. Hatakeyama's research papers are available in Tohoku University's repository at: https://cir.nii.ac.jp/all?q=Tohoku%20psychologica%20folia. Until now, they have not been widely disseminated and they warrant closer attention. The original purpose was to determine, through the use of VMI, whether Y.K. could fill a empty circle on a small, neutral card spontaneously with mentally imaged colour. The author used what later became known as the "Open Circle Test", which is described as follows:

> *One of the twenty subjects, YK, could see inside of the circle vivid images of concrete shape and colour, not mere colour image, at any time when a certain colour name was given to her. She gave a signal by tapping the desk with her right-hand forefinger, every time such an eidetic image appeared or it changed in its figure. She could also scan the images as if she were looking at paintings or photographs.* [42].

The investigator used a white card (12.5 cm by 20 cm) and a grey card (10 cm by 15 cm). A circle of 5 cm in diameter was drawn in the centre of the white card, and one of 2 cm on the grey one. The investigator asked the participant to imagine a blue colour inside the circle, and then red, yellow and green. With the white card first and then the grey card, 3 min were assigned to each of the four colours (eight tests in total). After each occasion, Y.K. was required to state whether a colour image appeared, and if it did, what it looked like. They were requested to draw an image if they saw one, on a sheet of paper with a pencil.

Two months after the first laboratory visit, Y.K. came to two more sessions using the same methods. Twenty-four tests were carried out. The findings showed that, at each test, Y.K. could project images associated with the appointed colour names; the VMI content changed over the 3 min of the test. Y.K. tapped the desk with her forefinger to indicate that an image had begun to appear inside the circle or if it transformed its shape. Y.K. produced drawings afterwards, and her responses to nine of the twenty-four tests are reproduced in Figure 3.

Hatakeyama's conclusion was that eidetic imagery is a constructive process similar to those of the other VMI kinds of memory and imagination imagery. Hatakeyama (1975) [43] stated the following:

> *Various experimental evidences given by a twenty-year-old female of extremely high eidetic ability shows that eidetic imagery is possessed of the constructive character. The word "constructive" means not only the fact that different features from the original stimulus appear in the image, but also, more positively speaking, that eidetic imagery has an aspect of being constructed like any other ordinary memory imagery. In this regard, it deserves emphasis that an eidetic image belongs to "imagery" and that it should not be construed merely as a photographic image, that is, an accurate copy of the original stimulus, or as a photographic memory, that is, an accurate retention of the original stimulus in visual memory.*

Thus, eidetic imagery appears to be a highly flexible and fluid form of VMI that does not necessarily require an evoking stimulus. Hatakeyama [42,43] divided his observations into five classes.

3.3.1. Spontaneous Eidetic Imagery Not Based on a Stimulus Presentation

As noted, Y.K. visualized vivid eidetic images inside the circle at any time colour names were given to her. There were some cases when the content of a produced image could not be identified, or when part of the image was discrepant with the original object that had been indicated (Figure 3). When Y.K. was asked to visualize "a small concrete object" on a projection screen (50 cm by 40 cm, placed 50 cm in front of her eyes), the image showed alterations in certain parts. Details of the objects were found to be incorrect, e.g., the image of a compact, the inside mirror, which could not be seen from the outside, was visualized, but the shape of the mirror was round, whereas her own compact had a square mirror.

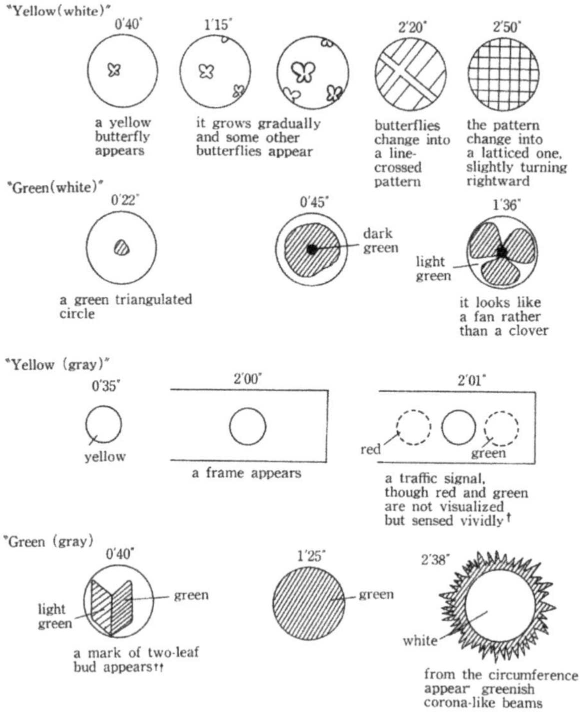

Figure 3. Y.K.'s original drawings in response to the Open Circle Test, with colouration as indicated [32]. "Blue," "Red," "Yellow," and "Green" correspond to the appointed colours; "white" and "grey" to the cards; numerals to the passage of time within the assigned 3 min. "†: At this time of the test, Y.K. was going to a drivers' school every day to get a license. However, the positions of "red" and "green" lamps in the imagined signal are contrary to the originals. ††: This mark must be placed on an automobile by beginner drivers in Japan. In the image, the yellow leaf of the original mark (the left-hand leaf) is altered to appear greenish. The drawings were produced by Y.K. The colours correspond to the participant's descriptions and are indicative only." Reproduced from [42].

3.3.2. Eidetic Imagery Elicited from Stimulus Presentation

Observations in [43] show highly constructive, sometimes inaccurate, characteristics of eidetic imagery, which included the following:

(i) Gradual Development and Fading;
(ii) Attentional Effort;
(iii) Supplementation: "new features, which the stimulus pictures did not contain, were added to the images";

(iv) Grounding: "where images lacked a particular part, such part was not left entirely blank but suffused as some kind of ground in the image";
(v) Metamorphosis: dramatic and sudden changes in eidetic imagery content;
(vi) Re-emergence;
(vii) Replication: e.g., "an orange square was presented and the same colour image appeared";
(viii) Figure–Ground Colour Reversal.

3.3.3. Eidetic Imagery Reproduced after a Lapse of Time

Eidetic images reproduced after a lapse of time revealed an "exceedingly constructive character". It took time before the eidetic images were produced. These images tended to appear "little by little". Details of the image were often initially indistinct and became more distinct. There were instances where motion or activity appeared in the image, e.g., "in the image of one silhouette picture evoked on the following day, a flag on the top of the building in the original picture, was seen fluttering". When the picture of the "Sparrows' School" was evoked on the following day, "the sparrows were moving about inside the nest, bobbing their heads; at the same time S sensed them peeping". There were some cases of divergence between the eidetic image and the memory contents of the original stimulus picture.

3.3.4. Eidetic Imagery Projected on the Surfaces with Complicated Patterns

When the projection cards presented complex designs, such as patterns of chiyogami (rice-paper with coloured figures) or a black-and-white striped pattern, eidetic images were constructed to assimilate elements of the projection surface.

3.3.5. Eidetic Imagery Evoked from Complex and Abstract Stimulus Figures

Tests with complex, abstract stimuli of Chinese characters used complicated numeral patterns (e.g., "twenty numerals drawn in blue, red, yellow or green, and scattered in various directions"). The eidetic imagery here were far from complete reproductions of the original stimuli, with "extremely constructive features", which lacked unity with other parts. The "entire image did not appear at the same time, but a restricted part, to which YK directed her eyes on the projection surface, was visualized and, as she shifted her eyes, the missing part appeared".

Hatakeyama's research [42,43] led to some interestingly novel conclusions:

(i) EI is a constructive process of agency involving the participant's attentional efforts, force of will and intentions. This point is reminiscent of Fechner's and Galton's findings;
(ii) EI is a flexible and fluid process with gradual development and fading, attentional effort, supplementation, grounding, metamorphosis and figure–ground colour reversal;
(iii) As suggested by Fechner, there is a dynamic interaction between the after-image and the eidetic image.

This study with a young woman of high eidetic imagery ability indicates the pinnacle of eidetic imagery achievement. Further studies by Hatakeyama [43] in different student age groups replicated Fechner's earlier-reported finding that after-images, following Emmert's Law, could be induced with eidetic imagery in one or more 20–24-year-olds, and more easily in children. To gain a realistic impression of eidetic imagery ability in the normal range, a study with a representative sample of the adult general population was necessary, as described next.

3.4. *Replication of Hatakeyama's Findings*

Two replication studies of Hatakeyama's findings were conducted by the author and Peter McKellar at the University of Otago, New Zealand [44]. Initially, we tested two eidetikers at the high end of the ability range. Our data replicated, in every detail, Hatakeyama's findings with Y.K. [42,43] (Figure 4). All of the characteristics previously observed by Hatakeyama [42,43] were repeated: gradual development and fading, attentional effort, supplementation, assimilation, grounding, fragmentation, metamorphosis and figure–ground colour reversal.

Next, we presented the Open Circle Test to randomly selected samples from the student population. Thirty of forty-four (68.2%) unselected students gave positive eidetic imagery reports in response to specific colour names [44]. Using less stringent criteria for eidetic imagery (e.g., the ability to see VMI projected in clouds or wallpaper patterns), eidetic imagery was reported by 100% of the sample. A strong relationship was obtained between the two types of eidetic imagery, "typographic" vs. "spontaneous" eidetic imagery, using the Easel Test and the Open Circle Test, respectively, in which all participants who could produce the former also produced the latter. Another replication by [45] with 327 students also showed a significant relationship between "typographic" eidetic imagery in the Easel Test and "voluntary or spontaneous" eidetic imagery, suggesting that these two kinds of eidetic imagery are essentially on a continuum of common ability.

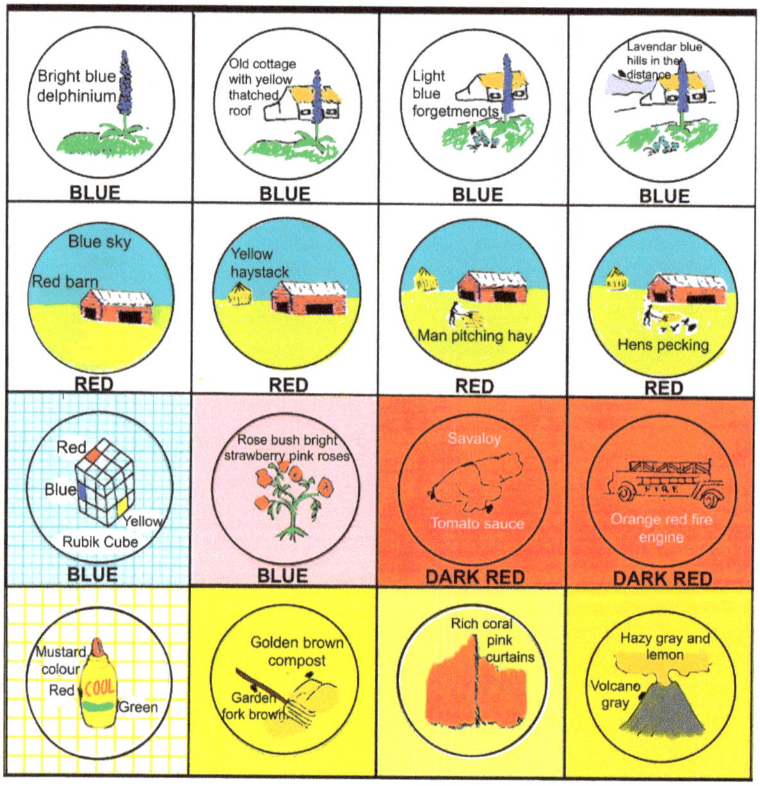

Figure 4. A high-ability participant's eidetic responses to ten trials with the Open Circle Test. Original black and white drawings were produced by the participant, with colouration as indicated. The top two rows show images evoked in each of two trials with a single colour name (trial 1:blue, and trial 2: red) over a five-minute period. The bottom two rows show responses for each of eight trials with varying papers and colours. The paper colour is shown (top two rows, white) and the evoking stimulus colour name, which was presented aurally, is printed beneath each circle. For the two tests using the dark red paper, the paper was stippled (stippling unshown) and elements of the produced images were assimilated with the stippling (replicating Hatakeyama's finding, Section 3.3.4). For the four trials in the bottom row, no colour name was presented, and the responses were evoked simply by the participant looking at the circle printed on four differently coloured papers. Notice how the four papers of varying shades of yellow spontaneously produced a diverse set of VMI [44].

Studies of eidetic imagery using unselected student samples suggest these conclusions:

(i) Eidetic imagery is a vivid, fluid and flexible process;
(ii) Eidetic imagery is a constructive process, not a reproductive or "photographic" one;
(iii) Eidetic imagery can be voluntarily controlled by the participant's focus of attention;
(iv) Like all other kinds of VMI, there are wide individual differences;
(v) In its weakest, spontaneous form, eidetic imagery is available to the entire population.

4. The Recent Period: 2000–2023

Mental Imagery in People with Intellectual Disabilities

There are only a few studies of mental imagery in people with intellectual disabilities. In the recent period, there have been efforts to fill this gap. Brown and Bullitis [46] explored mental imagery in 16 adults with intellectual disability (ID) in comparison to a group of 10 students who were seen individually in a "bare, windowless room". Photographs were briefly presented showing the following objects: (1) a tent, (2) a fire in a hearth, (3) a garden with a pool, (4) a woman doing a limbo exercise, (5) a man blowing out candles on a cake, (6) a woman by a step-ladder with a cup in her hand, (7) a chef cutting food, (8) a boat on a lake and (9) hikers in the mountains. Following the removal of the material, each individual was asked to describe any mental imagery that was evoked by the pictures. All respondents were video-taped, and their verbal responses were analysed. The ability to manipulate imagery was found to differ significantly between individuals and groups. However, all participants in both groups described a visual image spontaneously or when "provoked", and colour was spontaneously described by 15/16 ID individuals and by only 4/10 controls. A majority of both groups reported multi-sensory imagery and displayed signs of emotion and of movement.

Another new dataset has been provided by Hewitt et al. [47], who used participant-created drawings and semi-structured interviews to study the phenomenology of mental imagery in an ID group. The participants were ten people, two men and eight women aged from 25 to 54 (mean age = 52.3 years), who were assessed as having mild-to-moderate intellectual disabilities. They were invited to create an image, by drawing and/or by making a collage, to represent their experience of mental imagery. Participants were guided by the investigator to give a description of the image in its different parts and describe its overall meaning and the process of creating it. The *Vividness of Visual Imagery Questionnaire* [48] was presented together with prompts to explore the vividness levels of their mental imagery. The interviews took between 48 and 93 min (M = 62 min). The transcripts and images were analysed using interpretative phenomenological analysis [49]. The results are shown in Figure 5.

Hewitt et al.'s study suggests the following conclusions:

(i) People with ID experience and engage with mental imagery across different sensory modalities;
(ii) The clarity, detail and richness of the images is notable. As well as creating mental images, participants described changing and manipulating these images with VMI metamorphosis as observed by previous investigators described above;
(iii) Images could be "surprising, extraordinary and humorous";
(iv) Almost all participants described feelings associated with mental imagery, which "could be strong and powerful and encompassed a wide range of different emotions, such as surprise, delight, disgust and fear";
(v) There is a continuum of "mastery over mental imagery", with wide individual differences, as observed in all of the VMI studies reviewed here.

Another recent study [50] found no significant differences in VMI vividness or eidetic imagery ability in people with intellectual disability compared to typically developing individuals.

Superordinate themes	Subordinate themes
Theme 1: Mental imagery involves making, comparing, checking and changing images	Images can be clear and detailed Comparing different parts of an image Checking the details of an image Changing parts of an image Mental images can move
Theme 2: Our mental imagery can happen in any sense	Seeing mental images Touching mental images Hearing mental images Smelling mental images Tasting mental images Mental imagery can happen in more than one sense at the same time
Theme 3: Anything is possible in the world of mental imagery	Mental imagery can be surprising and funny Mental images can do impossible things vs mental images follow everyday rules We can share our mental images with other people
Theme 4: Mental imagery can have a big effect on our feelings	Mental imagery can give us strong feelings Mental imagery can make us feel lots of different feelings It can be easier or harder to imagine things that are important to us
Theme 5: Doing mental imagery can feel easy or difficult	Feeling like we're really good at mental imagery Sometimes mental imagery can feel too hard Asking us questions helps us have detailed mental imagery Helping us to concentrate makes mental imagery easier

Figure 5. Themes and subthemes identified by Hewitt et al. ([47], Creative Commons).

5. Synthesis of Findings

This review summarizes the findings of a specialized array of phenomenological investigations of VMI covering the entire history of mental imagery research. The review has collated, into a single analysis, a unique database that is drawn from a rich diversity of participants, beginning with Gustav Fechner (1860); then Galton's (1880) 100 distinguished "men of science'; samples of US university students; German, Japanese and New Zealand eidetic imagers; Africans in five countries on that continent; and groups of contemporary Britons with intellectual disabilities. In each and every group, there existed a widespread variability in the nature and vividness of VMI, but there is no obvious distinction in overall VMI abilities and characteristics between the different participant groups. The reviewed findings indicate an extraordinarily high level of consistency, replicability and coherence in phenomena that were written off for half a century by the behavioural school of psychology as "unscientific".

The purpose remains to synthesize the present findings in the form of a template that specifies the properties of VMI in healthy adult humans. The findings have comprised quantitative and qualitative data of a verbal and graphic nature on VMI phenomenology. This synthesis is inductive in nature, while taking into account the frequency of each of a complex of 16 characteristics across the ten datasets. The entire corpus of findings is synthesized in Figure 6. The letters "A, B, C, D. . . " label the 16 features to enable cross-referencing between Figure 6 and Table 4.

It can be seen that the template of VMI characteristics ranges from the most general at the top of the hierarchy ("Vividness") to the most specific at the bottom of the hierarchy ("Outline", "Detail", "Animation", "Feeling", etc.). To guide the construction of the template, it was necessary to weigh the relative importance of any descriptive term employed in the VMI phenomenological accounts using the frequency of each term as a proxy measure of a term's relevance and prominence.

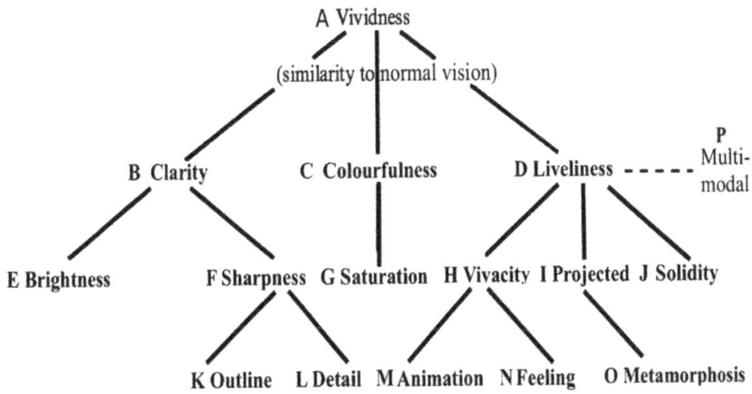

Figure 6. Taxonomy of VMI vividness based on the reviewed phenomenological studies. This hierarchical structure has four levels, with each term falling above and below its subordinate and superordinate constructs, respectively. The letters A–P indicate 16 interrelated characteristics that are tabulated across studies in Table 4.

Table 4. Synthesis of the findings of ten primary studies of VMI phenomenology. Columns A–O correspond to the 16 VMI features in the template of Figure 6.

Study	A	B	C	D	E	F	G	H	I	J	K	L	M	N	O	P
Fechner [21]	✓	✓	✓	✓	✓	✓	✓	✓	✓	✓	✓	✓	✓	-	✓	✓
Galton [24]	✓	✓	✓	✓	✓	✓	✓	✓	✓	✓	✓	✓	✓	-	-	✓
Armstrong [27]	✓	✓	✓	✓	✓	✓	✓	✓	✓	✓	✓	✓	✓	-	-	✓
Fernald [28]	✓	✓	✓	✓	✓	✓	✓	✓	✓	✓	✓	✓	✓	✓	✓	✓
Haber [36,37]	✓	✓	✓	✓	✓	✓	✓	✓	✓	✓	✓	✓	✓	✓	✓	✓
Doob [41]	✓	✓	✓	✓	✓	✓	✓	✓	✓	✓	✓	✓	✓	✓	✓	✓
Hatakeyama [42,43]	✓	✓	✓	✓	✓	✓	✓	✓	✓	✓	✓	✓	✓	✓	✓	✓
Marks and McKellar [44]	✓	✓	✓	✓	✓	✓	✓	✓	✓	✓	✓	✓	✓	✓	✓	✓
Brown and Bullitis [46]	✓	✓	✓	✓	✓	✓	✓	✓	✓	✓	✓	✓	✓	✓	✓	✓
Hewitt et al. [47]	✓	✓	✓	✓	✓	✓	✓	✓	✓	✓	✓	✓	✓	✓	✓	✓
TOTALS	10	10	10	10	10	10	10	10	10	10	10	10	10	7	8	10

The most frequently mentioned attribute of vivid VMI in the research participants' descriptions across the ten reviewed studies has been "colour", and the term "Colourfulness" is therefore included in the second tier of the template hierarchy, together with "Clarity" and "Liveliness".

The taxonomic rank at each level has been applied to the phenomenological descriptors of VMI vividness, according to their frequencies in the datasets in combination with their conceptual similarities and differences. The taxonomy started by establishing the taxonomic ranks in broad categories like "Low Vividness," "Moderate Vividness," and "High Vividness", and then proceeded with defining a second tier of specificity: "Clarity", "Colourfulness" and "Liveliness", and then a third and a fourth tier in a hierarchy.

Next, the content of the data from the reviewed studies suggested many other characteristics for classification that distinguish one level of vividness from another, including "Brightness", "Sharpness", "Saturation", "Vivacity", "Projected" and "Solidity". As noted, the categories are arranged hierarchically, with the most general one at the top ("Vividness") and more specific subcategories below (i.e., "Clarity", "Colourfulness", "Liveliness"). The fourth row indicates the most specific level of VMI characteristics: "Outline", "Detail", "Animation", "Feeling" and "Metamorphosis".

Table 4 indicates the extraordinarily high degree of consistency of findings across studies. A check mark in a cell indicates that the row study reported the column characteristic. A total of 155/160 cells are checked meaning that one indicator (N, Feeling) was unreported in three of the studies, and a second indicator (O, Metamorphosis) was unreported in two of the studies. These investigators did not solicit the necessary information about participants' emotional feelings or metamorphosis, and these five cells remain blank.

6. Discussion

The present review examines, collates and synthesizes historical and contemporary datasets on the phenomenology of VMI. As suggested in the Introduction, the process of discovering the nature of VMI is reminiscent of the parable of the blind monks inspecting the different part of an elephant (Figure 1). The current research findings are unique for their level of detail, scope, breadth and depth of the reviewed literature. The end product is a template defining VMI vividness with 16 descriptors.

Until now, our understanding of VMI has been fragmentary and piecemeal. The VMI template is a step towards describing the VMI "elephant" in its entirety for the first time. In light of this synthesis, it is possible to provide as comprehensive a specification of VMI vividness as possible using extant data. Previous definitions of VMI vividness had included clarity and liveliness, but not colourfulness [51,52]. However, the prominence of terms referring to "colour" in every one of the ten reviewed studies require the inclusion of colourfulness as an essential component of vividness. Notably, the principal instrument used for the assessment of VMI vividness, *The Vividness of Visual Imagery Questionnaire* (VVIQ), refers to colour in eight of its sixteen items [53].

The historical knowledge vacuum left by the neglect of VMI phenomenology led to a variety of incomplete or erroneous theoretical speculations and inadequate clinical applications. To give an example, the "Imagery Debate" [8,9] embraced two adversarial positions about the nature of VMI that were boldly expressed yet both were notably inconsistent with VMI phenomenology revealed by the present synthesis of findings. The first position {8} argued that VMI involves an analogue representation that is viewed by the imager as a "picture in the mind" or "mental photograph". The second position [9] argued that VMI involves tacit knowledge of the imaged situation, which causes the performance to proceed as the imager believes it should proceed if carried out perceptually. Neither the picture theory nor the tacit knowledge theory is consistent with the present findings. VMI *consists of a highly constructive and dynamic process that is both enactive and affective, having its own momentum, and often leading to unpredictable outcomes that can include the metamorphosis of the initial image content.* Thus, the present findings on VMI phenomenology are consistent with an enactive approach to perception and mental imagery, such as one proposed by this author [15,52].

Another example of a mismatch between assumptions and practice occurs in research attempting to link VMI content with the clinical condition of depression. Emily A. Holmes et al. [53] suggest that "individuals with depression may have particular difficulties with generating vivid and compelling imagery of positive events. Negative memories and images, in contrast, may come to mind all too readily" [53], i.e., depressed people experience depressing mental imagery. It is possible to escape the obvious tautology of this theory by exploring the links between VMI attributes such as colourfulness and depressive tendencies. For example, evoking colour imagery to determine whether people have a depressive tendency, the three colours of yellowish red, purple and dark grey are found to be significant discriminant variables [54]. These findings are consistent with the observation that Instagram photos posted by depressed individuals tend to be bluer, darker and greyer [55]. Using the VMI template in Figure 6, it is possible to make other predictions about the possible links between VMI and depressive mental states. For example, if the VMI of depressive people is explored over a period of minutes in the Open Circle Test, it can be predicted that they will observe image metamorphosis towards blue, dark and grey colours, towards less animation and lower detail compared to non-depressive people.

The template of Figure 6 can also be applied in future analyses of mental image vividness, which could lead to an improved understanding of how the 16 different vividness attributes contribute to the phenomenology of mental imagery. The current definition of "aphantasia", the alleged congenital absence of voluntary VMI defined by low scores on the VVIQ [56], could be more accurately explored by using all 16 of the template characteristics rather than simply VVIQ overall vividness total scores alone. It appears likely that significant numbers of people currently "diagnosed" as "aphantasic" would show positive responses to the Open Circle Test, with all 16 of the characteristics that are available for the assessment. This form of testing appears likely to reveal that many (or all) so-called "diagnosed aphantasics" are misdiagnosed false positives due to the inadequate method used for "diagnosis" [57].

By applying the taxonomic rank to the 16 criteria of phenomenological experiences of mental image vividness, future investigators will be able to create a structured framework that enhances our understanding of mental imagery experiences and contributes to the broader fields of cognitive science, vision science and psychology more generally.

7. Strengths and Limitations

The main strengths of this review are: (i) that it integrates for the first time findings from a set of nine largely neglected studies together with a widely-cited, but misinterpreted, study by Galton; (ii) that the selected studies represent highly diverse samples from five continents and 163 years of controlled investigation; and (iii) the impressive degree of consistency between these ten sets of findings. The main limitations are: (i) the relative lack of studies using non-verbal measures of VMI and (ii) the omission of publications in languages other than English, German and French.

8. Conclusions

This review and synthesis of diverse VMI phenomenological studies in healthy adults serves as a unique resource for investigators of individual differences, cognitive development and clinical and neurological conditions. The review has revealed an extraordinary degree of consistency and replicability of VMI characteristics across time, space and culture. VMI is a constructive process, with a continuum of intensity or vividness as its most prominent and frequently reported feature. The discovered template specifies the properties of VMI in healthy adult humans, taking into account the weight of evidence drawn from ten primary studies. Vividness is defined as a combination of clarity, colourfulness and liveliness, where clarity consists of brightness and sharpness, colourfulness consists of saturation, and liveliness consists of vivacity, animation, feeling, solidity, projection in three-dimensional space and metamorphosis.

The template provides a schema for further research on the phenomenological properties of VMI. The taxonomic template for the phenomenological experience of mental image vividness enhances our understanding of VMI experience and contributes to the broader fields of cognitive science, vision science and psychology more generally.

Funding: This research received no external funding.

Institutional Review Board Statement: Not applicable.

Informed Consent Statement: Not applicable.

Data Availability Statement: Not applicable.

Conflicts of Interest: The author declares no conflict of interest.

References

1. Tye, M. Qualia. Stanford Encyclopedia of Philosophy. 2021. Available online: https://plato.stanford.edu/entries/qualia/ (accessed on 14 October 2023).
2. Aristotle. "De Anima" (III, 7, 431a, 14–17). Available online: https://academic.oup.com/book/12823/chapter-abstract/163063885?redirectedFrom=fulltext (accessed on 14 October 2023).
3. Merleau-Ponty, M. *Phenomenology of Perception*; Smith, C., Translator; Routledge & Kegan Paul: Oxfordshire, UK, 1962.
4. Nanay, B. "Mental Imagery", The Stanford Encyclopedia of Philosophy (Winter 2021 Edition), Zalta, E.N., Ed. Available online: https://plato.stanford.edu/entries/mental-imagery/ (accessed on 14 October 2023).
5. Giorgi, A. Reflections on the Status and Direction of Psychology: An External Historical Perspective. *J. Phenomenol. Psychol.* **2013**, *44*, 244–261. [CrossRef]
6. Neisser, U. *Cognitive Psychology*; Appleton Century Crofts: New York, NY, USA, 1967.
7. Sheehan, P.W. (Ed.) *The Function and Nature of Imagery*; Academic Press: New York, NY, USA, 1972.
8. Kosslyn, S.M. The medium and the message in mental imagery: A theory. *Psychol. Rev.* **1981**, *88*, 46. [CrossRef]
9. Pylyshyn, Z.W. The imagery debate: Analogue media versus tacit knowledge. *Psychol. Rev.* **1981**, *88*, 16. [CrossRef]
10. Farah, M.J. The neurological basis of mental imagery: A componential analysis. *Cognition* **1984**, *18*, 245–272. [CrossRef] [PubMed]
11. Paivio, A. Mental imagery in associative learning and memory. *Psychol. Rev.* **1969**, *76*, 241–263. [CrossRef]
12. Shepard, R.N.; Metzler, J. Mental Rotation of Three-Dimensional Objects. *Science* **1971**, *171*, 701–703. [CrossRef] [PubMed]
13. McKelvie, S.J. The VVIQ and beyond: Vividness and its measurement. *J. Ment. Imag.* **1995**, *19*, 197–252.
14. Cumming, J.; Eaves, D.L. The Nature, Measurement, and Development of Imagery Ability. *Imagin. Cogn. Pers.* **2018**, *37*, 375–393. [CrossRef]
15. Marks, D.F. The Action Cycle Theory of Perception and Mental Imagery. *Vision* **2023**, *7*, 12. [CrossRef]
16. Richardson, A. Defining mental imagery. In *Mental Imagery*; Springer: Berlin/Heidelberg, Germany, 1969; pp. 1–12.
17. Binet, A. Mental imagery. *Fortnightly* **1892**, *52*, 95–104.
18. Charcot, J.M.; Bernard, D. Un cas de suppression brusque et isolée de la vision mentale des signes et des objets (formes et couleurs). *Le Progrès Médical* **1883**, *11*, 568–571.
19. Piaget, J.; Inhelder, B. Mental images. In *Experimental Psychology Its Scope and Method: Volume VII (Psychology Revivals): Intelligence*; Psychology Press: London, UK, 2014; p. 85.
20. Fechner, G. *Elements of Psychophysics (Volume 1) 1860*; Howes, D.H., Boring, E.G., Eds.; Adler, H.E., Translator; Holt, Rinehart and Winston: Washington, DC, USA, 1966.
21. Marks, D.F. Gustav Fechner's observations on his own and others' visual mental imagery. *J. Educ. Psychol. Res.* **2023**, *5*, 651–675.
22. Weiskrantz, L. An unusual case of after-imagery following fixation of an "imaginary" visual pattern. *Q. J. Exp. Psychol.* **1950**, *2*, 170–175. [CrossRef]
23. Kunzendorf, R.G. After-images of eidetic images: A developmental study. *J. Ment. Imag.* **1989**, *13*, 55–62.
24. Galton, F.I. Statistics of mental imagery. *Mind* **1880**, *5*, 301–318. [CrossRef]
25. Galton, F. *Inquiries into Human Faculty and Its Development*, 2nd ed.; Macmillan: New York, NY, USA, 1883; 1907 by J. M. Dent & Co. (Everyman) First Electronic Edition, 2001.
26. Brewer, W.F.; Schommer-Aikens, M. Scientists are not deficient in mental imagery: Galton revised. *Rev. Gen. Psychol.* **2006**, *10*, 130–146. [CrossRef]
27. Armstrong, A.C., Jr. The imagery of American students. *Psychol. Rev.* **1894**, *5*, 496. [CrossRef]
28. Fernald, M.R. The diagnosis of mental imagery. *Psychol. Monogr.* **1912**, *14*, i-169. [CrossRef]
29. Watson, J.B. Image and Affection in Behavior. *J. Philos. Psychol. Sci. Methods* **1913**, *10*, 421. [CrossRef]
30. McKellar, T.P.H. *Imagination and Thinking. A Psychological Analysis*; Cohen & West: London, UK, 1957.
31. Jaensch, E.R. *Über den Aufbau der Wahrnehmungswelt und die Grundlagen der Menschlichen Erkenntnis*, 2nd ed.; Barth: Leipzig, Germany, 1923.
32. Jaensch, E.R. *Eidetic Imagery and Typological Methods of Investigation*; Harcourt: New York, NY, USA, 1930.
33. Allport, G.W. Eidetic imagery. *Br. J. Psychol.* **1924**, *15*, 19–120. [CrossRef]
34. Klüver, H. Eidetic phenomena. *Psychol. Bull.* **1932**, *29*, 181–203. [CrossRef]
35. Gengerelli, J.A. Some Quantitative Experiments with Eidetic Imagery. *Am. J. Psychol.* **1930**, *42*, 399. [CrossRef]
36. Haber, R.N.; Haber, R.B. Eidetic imagery: I. Frequency. *Percept. Mot. Ski.* **1964**, *19*, 131–138. [CrossRef] [PubMed]
37. Haber, R.N. Twenty years of haunting eidetic imagery: Where's the ghost? *Behav. Brain Sci.* **1979**, *2*, 583–594. [CrossRef]
38. Photographic Memories Do Exist. The Harvard Crimson. Available online: https://www.thecrimson.com/article/1970/3/24/photographic-memories-do-exist-ptwo-harvard/ (accessed on 14 October 2023).
39. Stromeyer, C.F.; Psotka, J. The Detailed Texture of Eidetic Images. *Nature* **1970**, *225*, 346–349. [CrossRef] [PubMed]
40. Merritt, J.O. None in a million: Results of mass screening for eidetic ability using objective tests published in newspapers and magazines. *Behav. Brain Sci.* **1979**, *2*, 612. [CrossRef]
41. Doob, L.W. Eidetic Imagery: A Cross-Cultural Will-O'-The-Wisp? *J. Psychol.* **1966**, *63*, 13–34. [CrossRef]
42. Hatakeyama, T. The process of having identified an adult eidetic person and her eidetic experiences in everyday life. *Tohoku Psychol. Folia* **1974**, *33*, 102–118.

43. Hatakeyama, T. The constructive character of eidetic imagery. *Tohoku Psychol. Folia* **1975**, *34*, 38–51.
44. Marks, D.F.; McKellar, P. The nature and function of eidetic imagery. *J. Ment. Imag.* **1982**, *6*, 1–124.
45. Matsuoka, K.; Onizawa, T.; Hatakeyama, T. Incidence of young adult eidetikers, and two kinds of eidetic imagery. *Tohoku Psychol. Folia* **1987**, *46*, 62–74.
46. Brown, R.I.; Bullitis, E. The process of mental imagery in persons with or without intellectual disability: An exploratory project. *J. Dev. Disabil.* **2006**, *12*, 1–18.
47. Hewitt, O.M.; Langdon, P.E.; Hales, S.A.; Larkin, M. The phenomenology of mental imagery in people with intellectual disabilities. *Psychol. Psychother. Theory, Res. Pract.* **2022**, *96*, 25–39. [CrossRef] [PubMed]
48. Marks, D.F. Visual imagery differences in the recall of pictures. *Br. J. Psychol.* **1973**, *64*, 17–24. [CrossRef] [PubMed]
49. Smith, J.A.; Flowers, P.; Larkin, M. *Interpretative Phenomenological Analysis: Theory, Method and Research*; Sage: London, UK, 2021.
50. Syed, A.A.; Neelofur, S.; Moran, A.; O'Reilly, G. Exploring the vividness of mental imagery and eidetic imagery in people with intellectual disability (ID) in comparison with typically developing (TD) individuals. *Heliyon* **2020**, *6*, e05429. [CrossRef] [PubMed]
51. Marks, D.F. Individual differences in the vividness of visual imagery and their effect on function. In *The Function and Nature of Imagery*; Sheehan, P.W., Ed.; Academic Press: New York, NY, USA, 1972; pp. 83–108.
52. Marks, D.F. I Am Conscious, Therefore, I Am: Imagery, Affect, Action, and a General Theory of Behavior. *Brain Sci.* **2019**, *9*, 107. [CrossRef] [PubMed]
53. Holmes, E.A.; Blackwell, S.E.; Heyes, S.B.; Renner, F.; Raes, F. Mental imagery in depression: Phenomenology, potential mechanisms, and treatment implications. *Annu. Rev. Clin. Psychol.* **2016**, *12*, 249–280. [CrossRef] [PubMed]
54. Wu, F.-G.; Chang, E.; Lee, Y.-J. Manifestation of depressive tendency in color perception and colors utilized in creating a self-portrait. *Color Res. Appl.* **2008**, *34*, 84–92. [CrossRef]
55. Reece, A.G.; Danforth, C.M. Instagram photos reveal predictive markers of depression. *EPJ Data Sci.* **2017**, *6*, 15. [CrossRef]
56. Zeman, A.; Dewar, M.; Della Sala, S. Lives without imagery–Congenital aphantasia. *Cortex* **2015**, *73*, 378–380. [CrossRef]
57. Blomkvist, A.; Marks, D.F. Defining and 'diagnosing' aphantasia: Condition or individual difference? *Cortex* 2023, *in press*. [CrossRef]

vision

MDPI

Article

Associations between Autistic-like Traits and Imagery Ability

Takao Hatakeyama

Independent Researcher, Yamagata 990-8560, Japan; t_hatakeyama@cc.yamagata-u.ac.jp

Abstract: This article examines empirical associations between qualities of the imagination, mental imagery, and cognitive abilities with special reference to autism. This study is the first to explore the empirical relationships between autistic-like traits and tests of imagery differences. Imaginative impairments and distinctive sensory characteristics in individuals with autism spectrum disorder (ASD) should be reflected in their interactions with mental imagery. However, the relationship between ASD and imaging traits remains unclear. Based on the hypothesis that the degree of autistic-like traits is reflected in imagery traits, this study examined how the individual Autism Spectrum Quotient (AQ) relates to imagery ability in 250 college students. Two vividness tests and one imagery-type test were used to assess imagery ability. Scores in each imagery test were compared between the high-scoring group classified by the AQ and the rest of the participants and between the low-scoring group classified by the AQ and the other participants. This study also directly compared imagery test scores between the high- and low-scoring groups. In terms of the total AQ score, the high-scoring group exhibited lower visualization scores. Regarding AQ subscales, "imagination" had the most extensive relationship with imagery traits, with the high-scoring group (unimaginative) showing lower imagery vividness across various modalities as well as lower visualization and verbalization scores. This was followed by the "attention to detail" subscale, on which the high-scoring group (attentive to detail) showed higher vividness of visual imagery. The results of the low-scoring group exhibited, on the whole, opposite imagery tendencies to the high-scoring group. The results indicate that autistic-like traits are associated with qualities of the imagination and especially mental imagery ability.

Keywords: autism spectrum disorder (ASD); Autism Spectrum Quotient (AQ); imagery ability; vividness; imagery type; college students; Vividness of Visual Imagery Questionnaire (VVIQ); Questionnaire upon Mental Imagery (QMI); Verbalizer–Visualizer Questionnaire (VVQ)

1. Introduction

People image the world according to their individual neurocognitive traits and abilities. Here, we examine how imagery and imagination differences are associated with traits of autism spectrum disorder (ASD). ASD is characterized by persistent deficits in social communication and social interaction (e.g., deficits in non-verbal communicative behaviors and difficulties in imaginative play), as well as restricted and repetitive patterns of behavior, interests, or activities (e.g., hyper- or hyporeactivity to sensory input or unusual interest in sensory aspects of the environment) (DSM-5 [1]).

Deficits in non-verbal communication and difficulties in imaginative play are associated with imagination. For example, Scott and Baron-Cohen [2] asked children with autism to draw pictures of a real thing (e.g., a house, man, or spider) and something that does not exist (e.g., a house, man, or two-headed monster). Children with autism exhibited a specific impairment in their ability to imagine unusual or unreal objects, which was attributed to possible neural dissociation from other kinds of imagery [2].

Significantly, individuals with ASD exhibit traits such as hyper- or hyporeactivity to sensory input or unusual sensory interest. According to Tavassoli and Miller et al. [3], sensory processing is a key feature of individuals with ASD. Similarly, Kumagai [4] identified hypersensitivity as a core symptom of autism. Takahashi and Masubuchi [5] found that issues of hypersensitivity and insensitivity in people with Asperger's syndrome and high-functioning autism have not been sufficiently elucidated.

Sensory issues in individuals with and without ASD were investigated using questionnaires measuring sensory features. For example, Crane et al. [6] investigated the Adult/Adolescent Sensory Profile (AASP [7]) and found that sensory abnormalities are prevalent in ASD and that there is considerable within-group variability. Investigating the relationship between the Autism Spectrum Quotient (AQ) [8] and the AASP, Mayer [9] discovered that higher levels of autistic traits are associated with higher levels of sensory function atypicality in both neurotypical and ASD adults. Robertson and Simmons [10] developed the Glasgow Sensory Questionnaire (GSQ) to examine sensory problems and observed positive associations between the GSQ and the AQ, including its subscales. Takayama et al. [11] and Sapey-Triomphe et al. [12] reported similar results. Tavassoli and Miller et al. [3] investigated the relationship between ASD and the Sensory Over-Responsiveness (SensOR) subscale of the Sensory Processing Scale [13]. According to their results, adults with ASD report more sensory over-responsivity than controls across various sensory domains, and SensOR is positively correlated with AQ. Tavassoli and Hoekstra et al. [14] developed the Sensory Perception Quotient (SPQ), a questionnaire that investigates basic sensory functioning. Their findings revealed that adults with ASD report more sensory hypersensitivity to the SPQ than controls, and the SPQ is correlated with the AQ and SensOR. Several studies have indicated that in addition to the total AQ score, each subscale score predicts sensory functioning problems in daily life [9–12,15].

Imaginative impairment and distinctive sensory characteristics in individuals with ASD should be reflected in their interactions with imagery. However, it remains unclear how these traits are related to imagery traits. Kosslyn [16], based on his model, contended that imagery relies largely on perception mechanisms in the brain because visual images are generated, inspected, maintained, and transformed using the same processing mechanisms of visual perception. Research on the relationship between the AQ and sensory functioning problems allows us to predict the relationship between autistic traits and imagery characteristics.

This study examined the hypothesis that an individual's degree of "autistic-like traits", as measured by the AQ, will be reflected in their imagery ability. The AQ assumes cognitive and behavioral continuity among neurotypical individuals, those with Asperger's syndrome, and those with autism. Specifically, we explored the relationship between autistic-like traits and imagery abilities among college students. Participants were classified into high-, medium-, and low-scoring groups according to the quartile points of the total score and each subscale score of the Japanese version of the AQ [17]; that is, a higher quarter, a medium half, and a lower quarter. According to our prediction, the high-scoring group should exhibit a relatively high degree of autistic-like traits, whereas the low-scoring group should have the least autistic-like traits.

Regarding imagery ability, Hatakeyama classifies it into four dimensions: vividness, controllability, preference (imagery type), and absorption [18]. Large individual differences were revealed long ago by Galton in his study on vividness using the Breakfast Table Questionnaire [19]. Later, Betts developed the Questionnaire upon Mental Imagery (QMI), a 150-item seven-sensory-modality imagery vividness test [20] which Sheehan then shortened to 35 items, 5 for each modality [21]. Marks developed the Vividness of Visual Imagery Questionnaire (VVIQ), a 16-item vision-specific vividness test [22] which is widely used around the world to this day. Marks states: "By 'vividness' we mean a combination of clarity and liveliness. The more vivid an image, therefore, the closer it approximates an actual percept." [23] (p.83). Other tests of imagery ability include Gordon's Test of Visual Imagery Controllability (TVIC) [24]. This test assesses the ease with which a respondent is capable of imaging car scenes that are presented verbally in succession. Paivio's Individual Differences Questionnaire (IDQ) [25] and Richardson's Verbalizer–Visualizer Questionnaire (VVQ) [26] assess preference or imagery type. Imagery type refers to individual differences in imagery of sensory areas that predominate in learning, recall, thinking, and imagining.

The IDQ and VVQ focus on whether language or imagery is dominant. Absorption is a dimension of the degree of being absorbed in perceptual stimuli and scenes and fantasies or imaginations. A representative measure is Tellegen and Atkinson's Absorption Scale (AS) [27]. Although vividness has long been focused on as an imagery ability, it should be noted that vividness is not always necessary for other dimensions, and imagery in other dimensions does not need to be vivid. In this study, three imagery questionnaire tests, the VVIQ, a shortened version of the QMI, and the VVQ, were used to measure imagery ability.

We employed two kinds of analyses. First, to clarify what sort of imagery traits groups exhibit, we compared the scores of three imagery tests between the high-scoring group and the rest of the participants and between the low-scoring group and the other participants. Second, to identify differences between the high- and low-scoring groups, we directly compared the imagery test scores for these two extremes. If the results of the high-scoring group contrast with those of the low-scoring group in the first analysis, there is evidence that an individual's degree of autistic-like traits affects imagery ability. In such cases, a direct comparison between high- and low-scoring groups in the second analysis should show sharp differences. If only the high-scoring group shows relationships with imagery traits in the first analysis, such imagery traits could be recognized as belonging to individuals with a higher degree of autistic traits. However, if only the low-scoring group shows relationships with imagery traits in the first analysis, such imagery traits could be recognized as belonging to individuals with little or no autistic traits. In these cases, a direct comparison between the two groups does not necessarily reveal differences. In such a way, the first method of analysis can reveal not only the characteristics of the high-scoring group but also those of the low-scoring group.

2. Methods

2.1. Materials

Autistic-like traits were measured using the Japanese version [17] of the AQ [8], which comprises five subscales: social skill, attention switching, attention to detail, communication, and imagination.

We measured the imagery ability using two vividness tests and one imagery-type test. With respect to the former, we employed the Vividness of Visual Imagery Questionnaire (VVIQ, Marks [22], translated into Japanese by the author), which has four scenes (relative or friend, rising sun, familiar shop, and country scene), and the Questionnaire upon Mental Imagery (QMI, Richardson [28] and Sheehan [21], translated from [28] into Japanese by Onizawa and Takiura), which includes seven modalities (visual, auditory, cutaneous, kinesthetic, gustatory, olfactory, and organic). As an imagery-type test, we used the Verbalizer–Visualizer Questionnaire (VVQ, Richardson [26], translated into Japanese by the author), which consists of both verbal and visual items. However, for the VVQ, we did not adopt Richardson's [26] original scoring system, which collectively counts positive responses to visual items and negative responses to verbal items as visualizing scores. Instead, we adopted Hatakeyama's [29] scoring approach, which treats verbalization and visualization scores separately (VVQ-Verbal and VVQ-Visual, respectively).

2.2. Participants

The study sample comprised 250 students (125 males and 125 females; 18–23 years old, M = 18.9 years) from two colleges in Yamagata and Sendai, Japan. Based on gender, participants were classified into high-, medium-, and low-scoring groups according to the quartile points of the AQ total score and each subscale score (i.e., a higher quarter, a medium half, and a lower quarter). Classification by quartile points is appropriate because the 3rd quartile (75th percentile) point of the AQ total score almost corresponds to the cut-off point for Asperger's traits [30,31], though cut-off points with respect to the AQ subscales are not known. Participants were classified according to gender based on the finding that males scored higher than females on the AQ total score and subscales of social skill, attention switching, communication, and imagination in Baron-Cohen et al. [8].

Males also scored higher on the AQ total score and on the social skill, communication, and imagination subscales in Wakabayashi et al. [17]. By the classification according to gender, an analysis and discussion can be conducted without considering gender factors. The number of participants differed for each imagery test (VVIQ, $N = 250$; QMI, $N = 229$; VVQ, $N = 232$).

2.3. Procedure

The questionnaire survey was conducted in the classroom. On the cover of the questionnaire booklet, the purpose of the study, number of questions, data processing approach, and plan for presentations on meetings were provided. Each participant was required to record the date and their faculty and department, college, year, gender, and age but not their name. Prior to their participation in the study, all participants were informed orally about the purpose of the study, the voluntary nature of participation, anonymity in participation, data processing and research presentation, and the disclosure of information upon request from the participants. Verbal permission was obtained from the participants, and there was no harm. This study was approved by an ethics committee of a college.

2.4. Data Analyses

In the first analysis, the scores of the three imagery tests were analyzed using *t*-tests between the high-scoring group and a randomly selected third (25%) of the medium- and low-scoring combined group, and between the low-scoring group and a randomly selected third (25%) of the high- and medium-scoring combined group. The selection of one-third of participants other than the target (experimental) group (i.e., high- or low-scoring groups, respectively) was, for statistical convenience, to ensure that the size of the control group was similar to that of the experimental group. Every third participant was selected according to one of three routines. In the second analysis, direct comparisons were made using *t*-tests between the extremes of the continuum; that is, between the high- and low-scoring groups.

The second type of analysis, the extreme group approach (EGA), has been widely adopted in studies using the AQ, especially in the field of perception. However, as Preacher et al. [32] point out, there are costs associated with the EGA. An important issue is whether the lower-end group can be an appropriate control group when a large number of medium-scoring individuals are excluded from the spectrum. In this study, the first analysis was more appropriate because the control group comprised all participants except those in the higher- or lower-end groups of a continuum. However, as this was the first exploratory study to investigate the relationship between autistic-like traits and imagery traits, a second analysis comparing both end groups was conducted to elucidate the effect of autistic-like traits on imagery. The first analysis constitutes the primary analysis of this study, and the second is supplementary.

3. Results

3.1. Basic Statistics of the AQ

The AQ total score averaged 20.48 ($SD = 6.19$), somewhat higher than that of the general control and student groups in Baron-Cohen et al. [8] and similar to that of the general control and student groups in Wakabayashi et al. [17].

The percentage of the cut-off point, 26+ [30,31], was 19.6%, higher than both groups in Baron-Cohen et al. [8] and higher than the general control group and lower than the student group in Wakabayashi et al. [17].

Table 1 shows the correlation coefficients between the AQ total score and each of the subscale scores. The subscale attention to detail was unrelated to the other four subscale scores and had a low correlation with the AQ total score, though significant. So, this subscale does not contribute much to the AQ total score.

Table 1. Correlation coefficients between AQ total score and subscale scores.

	1	2	3	4	5	6
1 AQ (Total score)	-					
2 AQ (Social skill)	0.731 ***	-				
3 AQ (Attention switching)	0.685 ***	0.361 ***	-			
4 AQ (Attention to detail)	0.317 ***	−0.088	0.098	-		
5 AQ (Communication)	0.774 ***	0.523 ***	0.433 ***	0.064	-	
6 AQ (Imagination)	0.569 ***	0.333 ***	0.286 ***	−0.076	0.322 ***	-

Note: *** $p < 0.001$.

3.2. Groupings by AQ Score

Table 2 shows the mean AQ total and subscale scores by gender and the classification into high-, middle-, and low-scoring groups by the score, along with range and numbers. For the subscale imagination, males scored significantly higher than females ($t(248) = 3.05$, $p = 0.003$), and for the subscale attention switching, males scored marginally higher than females ($t(248) = 1.70$, $p = 0.090$). No gender differences were seen in the total score or the subscales social skill, attention to detail and communication, which was dissimilar to Baron-Cohen et al. [8] and Wakabayashi et al. [17].

Table 2. AQ total score and subscale score by gender and groupings by score.

	Gender				H		M		L	
		N	Mean (*SD*)	*p*	Range	*N* (%)	Range	*N* (%)	Range	*N* (%)
AQ (Total score)	Male	125	21.06 (6.42)		25–40	30 (24.0)	17–24	65 (52.0)	4–16	30 (24.0)
	Female	125	19.91 (5.92)		24–38	35 (28.0)	17–23	55 (44.0)	5–16	35 (28.0)
	Total	250	20.48 (6.19)			65 (26.0)		120 (48.0)		65 (26.0)
AQ (Social skill)	Male	125	3.78 (2.35)		6–9	29 (23.2)	3–5	56 (44.8)	0–2	40 (32.0)
	Female	125	3.56 (2.44)		6–10	25 (20.0)	2–5	74 (59.2)	0–1	26 (20.8)
	Total	250	3.67 (2.40)			54 (21.6)		130 (52.0)		66 (26.4)
AQ (Attention switching)	Male	125	5.33 (1.83)	†	7–9	31 (24.8)	5–6	57 (45.6)	0–4	37 (29.6)
	Female	125	4.94 (1.73)		7–9	26 (20.8)	4–6	70 (56.0)	1–3	29 (23.2)
	Total	250	5.14 (1.79)			57 (22.8)		127 (50.8)		66 (26.4)
AQ (Attention to detail)	Male	125	4.93 (2.11)		7–10	32 (25.6)	4–6	59 (47.2)	0–3	34 (27.2)
	Female	125	5.14 (1.87)		7–10	27 (21.6)	4–6	76 (60.8)	1–3	22 (17.6)
	Total	250	5.04 (1.99)			59 (23.6)		135 (54.0)		56 (22.4)
AQ (Communication)	Male	125	3.67 (2.17)		6–10	25 (20.0)	3–5	61 (48.8)	0–2	39 (31.2)
	Female	125	3.56 (2.02)		6–9	22 (17.6)	2–5	78 (62.4)	0–1	25 (20.0)
	Total	250	3.62 (2.09)			47 (18.8)		139 (55.6)		64 (25.6)
AQ (Imagination)	Male	125	3.34 (1.74)	**	5–8	29 (23.2)	3–4	54 (43.2)	0–2	42 (33.6)
	Female	125	2.70 (1.58)		4–9	32 (25.6)	2–3	64 (51.2)	0–1	29 (23.2)
	Total	250	3.02 (1.69)			61 (24.4)		118 (47.2)		71 (28.4)

(Note 1) H: high-scoring group (higher quarter); M: medium-scoring group (medium half); L: low-scoring group (lower quarter). (Note 2) ** $p < 0.01$, † $0.05 < p < 0.10$.

3.3. AQ Total Score and Imagery Tests

The results of both analyses are given below: first, the associations of the AQ total score with the imagery tests are provided; subsequently, for each imagery test, the associations of each AQ subscale are shown. As shown in Tables 3–6, we took note of the results that exhibited statistically significant and also marginally significant differences, revealing how the AQ relates to imagery tests without missing any possible results.

Table 3 exhibits the associations between the AQ total score and the imagery tests.

Table 3. Correlation coefficients between the AQ total score and imagery tests and a comparison of the mean scores (*SDs*) of the imagery tests for the groups classified by the AQ total score.

		H vs. M&L			H&M vs. L			H vs. L
	r	**H**	**M&L**	*p*	**H&M**	**L**	*p*	*p*
AQ (Total score)		*N* = 65	*N* = 61		*N* = 60	*N* = 65		
VVIQ (Relative/friend)	0.044	10.25 (3.69)	9.61 (3.42)		10.50 (3.51)	9.66 (3.48)		
VVIQ (Rising sun)	0.099	8.97 (2.86)	8.44 (3.52)		8.43 (3.29)	8.31 (3.57)		
VVIQ (Familiar shop)	−0.083	9.75 (3.63)	10.15 (3.40)		9.92 (3.19)	10.00 (3.41)		
VVIQ (Country scene)	0.020	10.74 (3.40)	10.82 (4.21)		9.92 (3.81)	10.68 (3.55)		
VVIQ (Total score)	0.027	39.71 (9.82)	39.02 (10.87)		38.77 (9.49)	38.65 (10.65)		
AQ (Total score)		*N* = 57	*N* = 57		*N* = 54	*N* = 62		
QMI (Visual)	0.046	12.47 (4.05)	12.36 (4.64)		12.63 (4.76)	12.25 (4.79)		
QMI (Auditory)	0.037	12.63 (4.54)	12.16 (4.41)		11.69 (4.80)	11.98 (4.19)		
QMI (Cutaneous)	0.069	14.25 (5.63)	14.30 (4.89)		13.50 (4.98)	13.15 (4.71)		
QMI (Kinesthetic)	0.078	12.33 (5.27)	11.49 (4.85)		11.63 (5.20)	11.19 (4.46)		
QMI (Gustatory)	0.061	11.46 (4.95)	10.74 (4.42)		10.54 (4.87)	10.77 (4.49)		
QMI (Olfactory)	0.082	15.53 (5.41)	15.05 (6.43)		14.52 (5.18)	*13.85* (5.13)		†
QMI (Organic)	−0.002	10.49 (4.55)	10.70 (5.34)		10.94 (5.35)	10.50 (4.82)		
QMI (Total score)	0.067	88.79 (25.38)	86.75 (26.03)		85.46 (27.31)	83.66 (25.32)		
AQ (Total score)		*N* = 59	*N* = 57		*N* = 54	*N* = 63		
VVQ-Verbal	−0.128 †	2.34 (1.28)	2.65 (1.47)		2.63 (1.63)	*2.83* (1.55)		†
VVQ-Visual	−0.162 *	5.03 (1.76)	**5.72** (1.73)	*	4.98 (1.62)	**5.83** (1.71)	**	*

(Note 1) H: high-scoring group (higher quarter); M: medium-scoring group (medium half); L: low-scoring group (lower quarter). (Note 2) Vividness of Visual Imagery Questionnaire (VVIQ); Questionnaire upon Mental Imagery (QMI): a lower score indicates higher vividness. (Note 3) Verbalizer–Visualizer Questionnaire (VVQ): Verbal = verbalization score; Visual = visualization score. (Note 4) ** $p < 0.01$, * $p < 0.05$, † $0.05 < p < 0.10$. Mean score in bold—significantly higher characteristic group; in italics—marginally higher characteristic group.

For the VVIQ, the AQ total score did not show any noticeable results in both analyses. For the QMI, in a comparison between both end groups, the high-scoring group (indicating more autistic-like traits) showed marginally weaker olfactory imagery than the low-scoring group ($t(117) = 1.73$, $p = 0.086$). For the VVQ, the AQ total score was related to imagery preference, with the high-scoring group showing significantly lower visualization scores than the medium- and low-scoring combined group ($t(114) = 2.11$, $p = 0.037$), while the low-scoring group (least autistic-like) showed higher visualization scores than the high- and medium-scoring combined group ($t(115) = 2.73$, $p = 0.007$). A direct comparison of the end groups revealed that the high-scoring group had significantly lower visualization scores than the low-scoring group ($t(120) = 2.52$, $p = 0.013$), and moreover, had marginally lower verbalization scores ($t(120) = 1.88$, $p = 0.062$).

3.4. AQ Subscales and the VVIQ

Table 4 exhibits the associations between the AQ subscales and the VVIQ.

Table 4. Correlation coefficients between the AQ subscales and the VVIQ and a comparison of the mean scores (*SDs*) of the VVIQ for the groups classified by the AQ subscale scores.

		H vs. M&L			H&M vs. L			H vs. L
	r	**H**	**M&L**	*p*	**H&M**	**L**	*p*	*p*
AQ (Social skill)		*N* = 54	*N* = 66		*N* = 63	*N* = 66		
VVIQ (Relative/friend)	0.173 **	10.52 (3.54)	*9.27* (3.43)	†	10.54 (3.41)	**8.94** (3.40)	**	*
VVIQ (Rising sun)	0.105 †	8.94 (3.04)	8.08 (3.34)		9.40 (3.58)	**7.70** (3.59)	**	*
VVIQ (Familiar shop)	−0.035	9.91 (3.37)	9.95 (3.32)		10.78 (3.30)	9.85 (3.50)		
VVIQ (Country scene)	0.037	10.02 (3.42)	10.18 (3.87)		11.43 (3.20)	**9.58** (4.16)	**	
VVIQ (Total score)	0.094	39.39 (9.95)	37.48 (9.91)		42.14 (9.01)	**36.06** (11.93)	***	

Table 4. *Cont.*

	r	H vs. M&L			H&M vs. L			H vs. L
		H	M&L	p	H&M	L	p	p
AQ (Attention switching)		$N = 57$	$N = 64$		$N = 62$	$N = 66$		
VVIQ (Relative/friend)	0.041	9.51 (3.75)	9.89 (3.48)		9.47 (3.95)	9.59 (3.37)		
VVIQ (Rising sun)	0.048	8.77 (3.15)	8.41 (3.66)		8.05 (3.33)	8.32 (3.73)		
VVIQ (Familiar shop)	−0.044	9.37 (3.58)	10.06 (3.74)		9.34 (3.51)	9.76 (3.24)		
VVIQ (Country scene)	−0.002	10.21 (4.15)	10.02 (3.79)		10.11 (3.93)	10.39 (3.66)		
VVIQ (Total score)	0.014	37.86 (11.04)	38.38 (11.89)		36.97 (11.58)	38.06 (9.27)		
AQ (Attention to detail)		$N = 59$	$N = 64$		$N = 64$	$N = 56$		
VVIQ (Relative/friend)	−0.237 ***	**8.81** (3.48)	10.09 (3.03)	*	**8.56** (3.40)	11.20 (3.18)	***	***
VVIQ (Rising sun)	−0.011	8.41 (3.67)	8.45 (3.22)		**7.63** (2.90)	8.79 (3.25)	*	
VVIQ (Familiar shop)	−0.108 †	***9.53*** (3.50)	10.55 (3.12)	†	10.02 (3.47)	11.04 (3.28)		*
VVIQ (Country scene)	−0.116 †	**9.81** (3.84)	10.66 (3.62)		**9.70** (4.31)	11.27 (3.62)	*	*
VVIQ (Total score)	−0.161 *	***36.56*** (10.26)	39.75 (8.82)	†	**35.91** (11.01)	42.29 (9.99)	***	**
AQ (Communication)		$N = 47$	$N = 67$		$N = 62$	$N = 64$		
VVIQ (Relative/friend)	0.027	9.51 (3.57)	9.28 (3.44)		10.00 (3.95)	9.25 (3.23)		
VVIQ (Rising sun)	0.009	8.23 (2.77)	8.16 (3.65)		8.39 (3.78)	7.86 (3.38)		
VVIQ (Familiar shop)	−0.080	**9.13** (3.41)	10.60 (3.77)	*	10.13 (3.51)	9.64 (3.38)		
VVIQ (Country scene)	0.015	10.21 (3.79)	10.15 (3.97)		10.19 (4.10)	9.86 (3.48)		
VVIQ (Total score)	−0.009	37.09 (9.74)	38.19 (10.96)		38.71 (11.71)	36.61 (9.64)		
AQ (Imagination)		$N = 61$	$N = 63$		$N = 59$	$N = 71$		
VVIQ (Relative/friend)	0.119 †	10.66 (3.61)	**9.27** (3.53)	*	10.75 (3.21)	**9.08** (3.48)	**	*
VVIQ (Rising sun)	0.165 **	9.48 (3.59)	**7.98** (3.66)	*	9.05 (3.56)	*8.25* (3.81)		†
VVIQ (Familiar shop)	0.018	10.41 (3.70)	10.35 (3.31)		11.46 (3.89)	**9.46** (2.93)	***	
VVIQ (Country scene)	0.140 *	11.36 (3.90)	**9.98** (3.71)	*	11.20 (4.22)	***9.89*** (3.58)	†	*
VVIQ (Total score)	0.151 *	41.90 (10.64)	**37.59** (11.03)	*	42.46 (11.27)	**36.69** (9.76)	**	**

(Note 1) H: high-scoring group (higher quarter); M: medium-scoring group (medium half); L: low-scoring group (lower quarter). (Note 2) Vividness of Visual Imagery Questionnaire (VVIQ): a lower score indicates higher vividness. (Note 3) *** $p < 0.001$, ** $p < 0.01$, * $p < 0.05$, † $0.05 < p < 0.10$. Mean score in bold—significantly higher characteristic group; in italics—marginally higher characteristic group; in italics with bold—marginally higher characteristic group but significantly between both ends.

3.4.1. Social Skill Subscale and the VVIQ

The high-scoring group (unsociable) showed marginally weaker imagery of a scene (i.e., a relative or friend) than the medium- and low-scoring combined group ($t(118) = 1.95$, $p = 0.053$). On the other hand, the low-scoring group (sociable) showed significantly higher vividness of visual imagery than the high- and medium-scoring combined group in three scenes (relative/friend, rising sun, and country) and the total VVIQ score (respectively, $t(127) = 2.67$, $p = 0.009$; $t(127) = 2.69$, $p = 0.008$; $t(127) = 2.83$, $p = 0.005$; $t(127) = 3.26$, $p = 0.001$). Between both end groups, the low-scoring group showed significantly higher vividness of imagery of two scenes (relative/friend and rising sun) (respectively, $t(118) = 2.49$, $p = 0.014$; $t(118) = 2.03$, $p = 0.045$).

3.4.2. Attention-Switching Subscale and the VVIQ

This subscale did not show any noticeable results with the VVIQ.

3.4.3. Attention-to-Detail Subscale and the VVIQ

The high-scoring group (attentive to detail) showed rather marginal but overall higher vividness of visual imagery than the medium- and low-scoring combined group. Specifically, the relative/friend scene showed significance ($t(121) = 2.18$, $p = 0.031$), whereas the familiar shop scene and total VVIQ score showed marginal significance (respectively, $t(121) = 1.71$, $p = 0.089$; $t(121) = 1.85$, $p = 0.066$). In contrast, the low-scoring group (inattentive to detail) showed significantly lower vividness of visual imagery than the high- and medium-scoring combined group in three scenes (relative/friend, rising sun, and country) and total VVIQ score (respectively, $t(118) = 4.36$, $p = 0.000$; $t(118) = 2.07$, $p = 0.041$; $t(118) = 2.14$, $p = 0.035$; $t(118) = 3.31$, $p = 0.001$). When comparing both end groups, the high-scoring group showed significantly higher vividness of visual imagery than the low-scoring group in three scenes (relative/friend, familiar shop, and country) and total VVIQ score (respectively, $t(113) = 3.83$, $p = 0.000$; $t(113) = 2.39$, $p = 0.019$; $t(113) = 2.09$, $p = 0.039$; $t(113) = 3.03$, $p = 0.003$).

3.4.4. Communication Subscale and the VVIQ

The high-scoring group (uncommunicative) showed significantly higher vividness of a visual scene (a familiar shop) than the medium- and low-scoring combined group ($t(112) = 2.13$, $p = 0.035$). For the low-scoring group (communicative), no noticeable results were observed. Between both end groups, no significances were observed.

3.4.5. Imagination Subscale and the VVIQ

The high-scoring and low-scoring groups exhibited close relationships with the vividness of visual imagery. More specifically, the high-scoring group (unimaginative) showed significantly lower imagery vividness than the medium- and low-scoring combined group for the three visual scenes (relative/friend, rising sun, and country) and total VVIQ score (respectively, $t(122) = 2.16$, $p = 0.032$; $t(122) = 2.29$, $p = 0.024$; $t(122) = 2.02$, $p = 0.046$; $t(122) = 2.22$, $p = 0.029$). In contrast, compared to the high- and medium-scoring combined group, the low-scoring group (imaginative) showed significantly higher vividness in the two visual scenes (relative/friend and familiar shop) and total VVIQ score (respectively, $t(128) = 2.81$, $p = 0.006$; $t(128) = 3.33$, $p = 0.001$; $t(128) = 3.13$, $p = 0.002$) and showed marginally higher vividness in a scene (country) ($t(128) = 1.92$, $p = 0.057$). Between both end groups, clear contrasts were seen for imagery vividness, with the high-scoring group (unimaginative) exhibiting significantly lower vividness in the two visual scenes (relative/friend and country) and total VVIQ score (respectively, $t(130) = 2.54$, $p = 0.012$; $t(130) = 2.26$, $p = 0.025$; $t(130) = 2.93$, $p = 0.004$) and marginally lower vividness in a scene (rising sun) ($t(130) = 1.89$, $p = 0.061$).

3.5. AQ Subscales and the QMI

Table 5 exhibits the associations between the AQ subscales and the QMI.

Table 5. Correlation coefficients between the AQ subscales and the QMI and a comparison of the mean scores (*SDs*) of the QMI for the groups classified by the AQ subscale scores.

		H vs. M&L			**H&M vs. L**			**H vs. L**
	r	**H**	**M&L**	*p*	**H&M**	**L**	*p*	*p*
AQ (Social skill)		*N* = 47	*N* = 63		*N* = 58	*N* = 62		
QMI (Visual)	0.113 †	12.64 (4.68)	12.31 (4.72)		12.72 (4.28)	*11.34* (4.48)	†	
QMI (Auditory)	0.036	11.43 (4.31)	12.51 (4.07)		12.33 (4.19)	11.50 (4.22)		
QMI (Cutaneous)	0.119 †	13.96 (5.25)	14.22 (4.70)		13.64 (5.14)	12.53 (4.45)		
QMI (Kinesthetic)	0.080	11.72 (5.45)	11.08 (4.36)		11.29 (4.52)	10.71 (4.51)		
QMI (Gustatory)	0.101	11.51 (5.00)	11.61 (4.96)		10.74 (4.75)	10.50 (4.46)		
QMI (Olfactory)	0.152 *	15.94 (5.99)	14.92 (5.18)		15.19 (5.78)	**13.47** (5.03)	†	*
QMI (Organic)	0.059	9.96 (4.95)	10.92 (4.74)		10.62 (4.22)	9.48 (4.60)		
QMI (Total score)	0.119 †	86.70 (26.38)	88.34 (25.34)		86.43 (25.23)	79.53 (25.13)		
AQ (Attention switching)		*N* = 52	*N* = 59		*N* = 55	*N* = 64		
QMI (Visual)	−0.011	12.00 (4.35)	12.63 (5.13)		11.93 (4.77)	12.53 (4.68)		
QMI (Auditory)	−0.022	11.60 (4.85)	12.02 (4.20)		12.05 (4.80)	12.48 (4.05)		
QMI (Cutaneous)	0.011	13.12 (5.65)	14.20 (5.35)		12.78 (5.15)	13.73 (4.48)		
QMI (Kinesthetic)	0.008	11.29 (4.84)	11.17 (5.15)		10.45 (4.17)	11.47 (4.55)		
QMI (Gustatory)	−0.032	10.62 (4.58)	11.32 (4.70)		10.65 (4.79)	11.34 (4.87)		
QMI (Olfactory)	0.034	14.40 (5.75)	15.46 (6.19)		14.69 (6.09)	14.30 (4.92)		
QMI (Organic)	−0.070	**9.35** (4.13)	11.29 (5.12)	*	10.38 (5.37)	11.22 (4.77)		*
QMI (Total score)	−0.015	82.31 (26.17)	87.83 (29.25)		82.40 (26.13)	87.08 (25.31)		

Table 5. *Cont.*

		H vs. M&L			H&M vs. L			H vs. L
	r	**H**	**M&L**	*p*	**H&M**	**L**	*p*	*p*
AQ (Attention to detail)		*N* = 55	*N* = 57		*N* = 59	*N* = 51		
QMI (Visual)	−0.118 †	*11.95* (4.42)	13.39 (4.69)	†	**11.29** (4.57)	13.60 (4.63)	**	†
QMI (Auditory)	−0.018	11.98 (4.58)	12.55 (3.93)		11.75 (4.50)	12.23 (4.79)		
QMI (Cutaneous)	−0.082	13.36 (5.83)	14.00 (4.25)		13.22 (5.02)	14.25 (4.91)		
QMI (Kinesthetic)	−0.067	11.38 (5.27)	12.14 (4.34)		10.95 (4.67)	12.14 (4.74)		
QMI (Gustatory)	−0.059	10.96 (5.49)	11.89 (5.04)		10.57 (4.81)	11.47 (4.24)		
QMI (Olfactory)	−0.092	14.85 (6.47)	14.75 (5.14)		*13.98* (5.23)	15.98 (5.66)	†	
QMI (Organic)	−0.112 †	10.20 (5.37)	11.30 (5.14)		10.00 (4.53)	11.41 (5.31)		
QMI (Total score)	−0.106	84.31 (29.51)	90.29 (24.42)		*82.42* (26.74)	91.08 (25.47)	†	
AQ (Communication)		*N* = 42	*N* = 62		*N* = 60	*N* = 61		
QMI (Visual)	0.063	12.69 (4.46)	11.41 (4.50)		12.07 (4.40)	11.73 (4.69)		
QMI (Auditory)	−0.005	12.10 (4.83)	11.67 (4.49)		11.83 (4.13)	11.90 (4.49)		
QMI (Cutaneous)	0.012	13.50 (5.58)	13.37 (5.54)		13.70 (5.37)	13.15 (4.36)		
QMI (Kinesthetic)	0.058	11.38 (4.56)	11.21 (5.04)		11.02 (4.96)	10.79 (4.33)		
QMI (Gustatory)	0.070	11.15 (4.65)	10.52 (4.77)		10.93 (4.46)	10.44 (4.33)		
QMI (Olfactory)	0.054	14.64 (5.15)	14.58 (6.44)		15.53 (6.17)	*13.72* (5.22)	†	
QMI (Organic)	0.010	10.38 (3.98)	10.29 (5.54)		11.08 (5.47)	9.84 (4.12)		
QMI (Total score)	0.056	86.63 (24.77)	82.90 (28.49)		86.39 (25.65)	81.51 (24.18)		
AQ (Imagination)		*N* = 57	*N* = 60		*N* = 55	*N* = 65		
QMI (Visual)	0.087	13.21 (4.11)	**11.17** (3.95)	**	12.82 (3.79)	12.18 (4.89)		
QMI (Auditory)	0.137 *	13.16 (4.59)	**11.32** (4.38)	*	12.60 (3.86)	**11.62** (4.00)		*
QMI (Cutaneous)	0.159 *	14.60 (4.87)	**12.77** (4.70)	*	15.07 (5.05)	**12.55** (4.47)	**	*
QMI (Kinesthetic)	0.176 **	12.25 (4.89)	11.30 (4.88)		12.95 (5.25)	**10.51** (3.99)	**	*
QMI (Gustatory)	0.103	11.51 (4.78)	10.39 (4.64)		11.89 (4.90)	*10.36* (4.50)	†	
QMI (Olfactory)	0.097	15.46 (4.73)	14.18 (5.85)		16.05 (5.57)	**13.65** (5.12)	*	*
QMI (Organic)	0.106	11.88 (5.06)	**9.90** (4.93)	*	11.36 (4.96)	*10.29* (4.40)		†
QMI (Total score)	0.155 *	91.81 (23.57)	**81.41** (24.38)	*	92.49 (24.96)	**81.52** (24.82)	*	*

(Note 1) H: high-scoring group (higher quarter); M: medium-scoring group (medium half); L: low-scoring group (lower quarter). (Note 2) Questionnaire upon Mental Imagery (QMI): a lower score indicates higher vividness. (Note 3) ** $p < 0.01$, * $p < 0.05$, † $0.05 < p < 0.10$. Mean score in bold—significantly higher characteristic group; in italics—marginally higher characteristic group; in italics with bold—marginally higher characteristic group but significant between both ends.

3.5.1. Social Skill Subscale and the QMI

The low-scoring group (sociable) showed marginally higher vividness of visual imagery and olfactory imagery (respectively, $t(118) = 1.73$, $p = 0.086$; $t(118) = 1.74$, $p = 0.084$). Between both end groups, the low-scoring group showed significantly higher vividness of olfactory imagery than the high-scoring group ($t(107) = 2.34$, $p = 0.021$).

3.5.2. Attention-Switching Subscale and the QMI

The high-scoring group (attention-unswitchable) showed significantly higher vividness of organic imagery than the medium- and low-scoring combined group ($t(109) = 2.18$, $p = 0.031$). For the low-scoring group (attention-switchable), no significant differences were observed. In a direct comparison of the two groups, the high-scoring group showed significantly higher vividness of organic imagery than the low-scoring group ($t(114) = 2.23$, $p = 0.027$).

3.5.3. Attention-to-Detail Subscale and the QMI

The high-scoring group (attentive to detail) showed marginally higher vividness of visual imagery than the medium- and low-scoring combined group ($t(110) = 1.67$, $p = 0.098$). In contrast, the low-scoring group (inattentive to detail) showed significantly lower vividness of visual imagery than the high- and medium-scoring combined group ($t(108) = 2.62$, $p = 0.010$), and marginally lower vividness of olfactory imagery, as well as of the whole modality (i.e., the total QMI score) (respectively, $t(108) = 1.92$, $p = 0.057$; $t(106) = 1.72$, $p = 0.089$). When comparing both end groups, the high-scoring group showed marginally higher vividness of visual imagery than the low-scoring group ($t(105) = 1.89$, $p = 0.062$).

3.5.4. Communication Subscale and the QMI

The high-scoring group (uncommunicative) showed no noticeable results. The low-scoring group (communicative) showed marginally more vivid olfactory imagery than the high- and medium-scoring combined group ($t(119) = 1.74$, $p = 0.084$).

3.5.5. Imagination Subscale and the QMI

The high-scoring group (unimaginative) showed significantly lower imagery vividness across the following modalities than the medium- and low-scoring combined group: visual, auditory, cutaneous, and organic imagery and total QMI score ($t(115) = 2.74$, $p = 0.007$; $t(115) = 2.22$, $p = 0.028$; $t(115) = 2.07$, $p = 0.041$; $t(115) = 2.14$, $p = 0.034$; $t(114) = 2.34$, $p = 0.021$). In contrast, compared to the high- and medium-scoring combined group, the low-scoring group (imaginative) showed significantly higher vividness across various modalities: cutaneous, kinesthetic, and olfactory imagery and total QMI score ($t(118) = 2.90$, $p = 0.004$; $t(118) = 2.89$, $p = 0.005$; $t(118) = 2.47$, $p = 0.015$; $t(117) = 2.40$, $p = 0.018$). In addition, the low-scoring group showed marginally higher vividness of gustatory imagery ($t(117) = 1.78$, $p = 0.078$). Between both end groups, clear contrasts were seen for imagery vividness across various modalities, with the high-scoring group (unimaginative) exhibiting significantly lower vividness in auditory, cutaneous, kinesthetic, and olfactory imagery and total QMI score ($t(120) = 1.98$, $p = 0.050$; $t(120) = 2.42$, $p = 0.017$; $t(120) = 2.16$, $p = 0.033$; $t(120) = 2.02$, $p = 0.046$; $t(119) = 2.33$, $p = 0.021$). Additionally, the high-scoring group showed marginally lower vividness in organic imagery ($t(120) = 1.85$, $p = 0.067$).

3.6. *AQ Subscales and the VVQ*

Table 6 exhibits the associations between the AQ subscales and the VVQ.

Table 6. Correlation coefficients between the AQ subscales and the VVQ and a comparison of the mean scores (*SDs*) of the VVQ for the groups classified by the AQ subscale scores.

		H vs. M&L			H&M vs. L			H vs. L
	r	**H**	**M&L**	*p*	**H&M**	**L**	*p*	*p*
AQ (Social skill)		*N* = 49	*N* = 63		*N* = 58	*N* = 60		
VVQ-Verbal	−0.121 †	2.33 (1.55)	2.65 (1.17)		2.53 (1.42)	**2.97** (1.53)		*
VVQ-Visual	−0.202 **	5.06 (1.76)	5.46 (1.76)		5.07 (1.84)	**5.85** (1.64)	*	*
AQ (Attention switching)		*N* = 51	*N* = 60		*N* = 57	*N* = 64		
VVQ-Verbal	−0.083	2.45 (1.32)	2.78 (1.39)		2.60 (1.43)	2.83 (1.45)		
VVQ-Visual	−0.071	5.10 (1.90)	5.42 (1.84)		5.61 (1.66)	5.38 (1.73)		
AQ (Attention to detail)		*N* = 55	*N* = 58		*N* = 59	*N* = 52		
VVQ-Verbal	0.045	2.69 (1.26)	2.43 (1.51)		2.47 (1.25)	2.52 (1.53)		
VVQ-Visual	0.152 *	*5.56* (1.78)	5.05 (1.91)		**5.59** (1.67)	4.94 (1.72)	*	†
AQ (Communication)		*N* = 43	*N* = 63		*N* = 59	*N* = 62		
VVQ-Verbal	−0.132 *	2.40 (1.43)	2.87 (1.49)		2.88 (1.46)	**3.18** (1.52)		**
VVQ-Visual	−0.092	4.93 (1.83)	*5.59* (1.75)	†	5.44 (1.87)	*5.61* (1.73)		†
AQ (Imagination)		*N* = 57	*N* = 59		*N* = 55	*N* = 65		
VVQ-Verbal	−0.105	2.35 (1.43)	**3.05** (1.51)	*	2.51 (1.51)	**3.09** (1.37)	*	**
VVQ-Visual	−0.306 ***	4.51 (1.68)	**5.92** (1.71)	***	5.02 (1.64)	**5.71** (1.67)	*	***

(Note 1) H: high-scoring group (higher quarter); M: medium-scoring group (medium half); L: low-scoring group (lower quarter). (Note 2) Verbalizer–Visualizer Questionnaire (VVQ): Verbal = verbalization score; Visual = visualization score. (Note 3) *** $p < 0.001$, ** $p < 0.01$, * $p < 0.05$, † $0.05 < p < 0.10$. Mean score in bold—significantly higher characteristic group; in italics—marginally higher characteristic group.

3.6.1. Social Skill Subscale and the VVQ

The high-scoring group (unsociable) did not show any noticeable results. On the other hand, the low-scoring group (sociable) showed significantly more visualization than the high- and medium-scoring combined group ($t(116) = 2.44$, $p = 0.016$). Between both end groups, the low-scoring group showed significantly more verbalization and visualization (respectively, $t(107) = 2.16$, $p = 0.033$; $t(107) = 2.42$, $p = 0.017$).

3.6.2. Attention-Switching Subscale and the VVQ

This subscale did not show any noticeable results with the VVQ.

3.6.3. Attention-to-Detail Subscale and the VVQ

The high-scoring group (attentive to detail) showed no noticeable results. The low-scoring group (inattentive to detail) showed significantly lower visualization than the high- and medium-scoring combined group ($t(109) = 2.02$, $p = 0.046$). When comparing both end groups, the high-scoring group showed marginally higher visualization than the low-scoring group ($t(105) = 1.83$, $p = 0.070$).

3.6.4. Communication Subscale and the VVQ

The high-scoring group (uncommunicative) showed marginally lower visualization scores than the medium- and low-scoring combined group ($t(104) = 1.87$, $p = 0.065$). The low-scoring group (communicative) did not show any noticeable results. Between both ends, the high-scoring group showed significantly lower verbalization and marginally lower visualization scores than the low-scoring group (respectively, $t(103) = 2.65$, $p = 0.009$; $t(103) = 1.94$, $p = 0.055$).

3.6.5. Imagination Subscale and the VVQ

The high-scoring and low-scoring groups exhibited close relationships between preference. More specifically, the high-scoring group (unimaginative) showed significantly lower verbalization and visualization scores than the medium- and low-scoring combined group (respectively, $t(114) = 2.56$, $p = 0.012$; $t(114) = 4.47$, $p = 0.000$). In contrast, compared to the high- and medium-scoring combined group, the low-scoring group (imaginative) showed higher verbalization and visualization scores (respectively, $t(118) = 2.22$, $p = 0.029$; $t(118) = 2.27$, $p = 0.025$). Between both end groups, clear contrasts were seen for preference. The high-scoring group showed significantly lower verbalization and visualization scores than the low-scoring group (respectively, $t(120) = 2.92$, $p = 0.004$; $t(120) = 3.94$, $p = 0.000$).

3.7. Correlation Coefficients between the AQ and Imagery Tests

Concerning correlation coefficients between the AQ and imagery tests, Tables 3–6 show that coefficients were generally small and, even when statistically significant, most were less than 0.2, with only three coefficients exceeding 0.2. Moreover, discrepancies between the coefficients and our comparative analyses were also revealed. For example, in the case of the AQ imagination subscale, the subscale correlated with VVQ-Visual and did not correlate with VVQ-Verbal, but in the comparisons for both associations, high- and low-scoring groups exhibited respectively clear qualities of visualization and verbalization. Accordingly, we will discuss the results of the comparative analyses in the following section.

4. Discussion

ASD is characterized by (1) persistent deficits in social communication and social interaction across multiple contexts and (2) restricted, repetitive patterns of behavior, interests, or activities (DSM-5 [1]). The former are exemplified by symptoms such as deficits in non-verbal communicative behaviors and difficulties in sharing imaginative play; meanwhile, the latter are exemplified by hyper- or hyporeactivity to sensory input or unusual interest in sensory aspects of the environment. Based on the hypothesis that an individual's degree of autistic-like traits is reflected in their imagery traits, this study examined how the AQ score is related to imagery test scores, particularly observing how imagery traits relate to both characteristics of ASD.

This section discusses the results that exhibited statistically significant and marginally significant differences in the first analysis and those that exhibited statistically significant differences in the second analysis because this study is in the exploratory stage. Doing so is necessary to avoid missing any possible results from the first analysis. Additionally, extreme group comparisons increase the odds of achieving statistical significance (Preacher et al. [32]), allowing for the exclusion of uncertain results in the second analysis.

For the AQ total score, the analyses revealed differences between the high- and low-scoring groups with respect to visualization. Briefly, individuals with a relatively high degree of autistic-like traits as a whole exhibited low visualization scores, whereas those with the fewest autistic traits had high visualization scores. This indicates that a higher degree of possession of autistic-like traits as a whole was related to impediments to visualization but was not related to vividness traits, so we cannot know anything about the vividness dimension for the AQ total score. Why was it not related to the vividness dimension? It is quite probable that setting off the effects of two traits measured by the subscales of the AQ—that is, the merit of "attention to detail" against the fault of "imagination" in imagery qualities, as presented below—resulted in no noticeable differences in the vividness dimension of imagery. If this was the case, the familiar method of classifying participants by their total AQ score would not provide sufficient knowledge of their imagery qualities.

On the AQ, the imagination subscale showed the most extensive relationship with imagery traits. More specifically, the high-scoring group (unimaginative) showed lower imagery vividness across various modalities and lower verbalization and visualization scores, whereas the low-scoring group (imaginative) showed higher imagery traits. A direct comparison revealed a clear contrast between the groups. These findings indicate that imagination is significantly associated with imagery traits. These close relationships may provide important clues for elucidating ASD. Moreover, the results demonstrate that imagination relates not only to imagery vividness and visualization but also to verbalization, suggesting that verbalization is more or less based on imagination.

We will look over recent studies regarding the effectiveness of the AQ subscales for perceptual characteristics of ASD, anticipating some possible bases for the results. Studies have shown that the imagination subscale of the AQ predicts several issues. Examples include reliance on high-spatial frequency features, that is, detail-focused processing bias (i.e., having an eye for detail) [33]; deficiency in multisensory integration due to a narrow temporal window [34]; and a diminished gaze-orienting effect for happy faces which is attributed to a lack of imagination when explored in experimental conditions using a computer monitor [35]. The first two characteristics, which are typified in detail-focused processing bias, may be the bases for poor imagination, whereas the third is the outcome.

Following imagination, the attention-to-detail subscale showed a strong relationship with imagery traits. The high-scoring group (attentive to detail) exhibited higher vividness of visual imagery, whereas the low-scoring group (inattentive to detail) showed lower vividness of visual, olfactory, and overall imagery and lower visualization scores. Direct comparisons between the two groups revealed that the attentive-to-detail group had higher vividness of visual imagery. The contrast between the two groups in terms of the vividness of visual imagery indicates that the degree of attention to detail first affects visual imagery vividness. Findings other than visual imagery were solely for the low-scoring group. Lower vividness and visualization are traits of people who are not detail-oriented.

Some studies have suggested that attention to detail plays a role in the vividness of imagery. Marks [36], in his review of historical datasets, showed that detail is one of the key components of visual imagery vividness. Richardson and Patterson [37] reported that multimodal sensory awareness training enhanced imagery vividness wherein participants observed objects minutely and subsequently formed images. Hishitani [38] found that developing expertise increases not only imagery vividness but also the controllability and frequency of imagery use. Skill expertise involves focusing on relevant skills and related events over an extended period. The present study's finding concerning vividness aligns with those of the aforementioned studies and supports the argument that "attention to the details of a stimulus" acts as a mechanism of imagery vividness (Hatakeyama [18]). ASD sensory traits, such as hyperactivity to sensory input or unusual interest in sensory aspects [1,3–5,10], can be categorized as attention to detail.

The attention-to-detail subscale of the AQ has been shown to predict several perceptual issues in ASD, such as superior performance on the Embedded Figure Test (EFT), which is understood as an advantage in processing detail [39], and a diminished gaze-orienting effect for happy faces; that is, a tendency to focus more on details and less on the whole [35]. These characteristics reflect the effects of attention on detailed traits. Alink and Charest [33] reported that among the AQ subscales, only the attention-to-detail subscale was not associated with the reliance-on-detail measure. For individuals with detailed traits, reliance on details may no longer be necessary.

For the social skill subscale, a result revealed a marginal association with the vividness of relative/friend imagery in the high-scoring group (unsociable), indicating that unsociable individuals may have weaker person imagery. Except for this, the social skill subscale exhibited relationships with imagery traits solely in the low-scoring group (sociable), whereby sociable individuals showed higher vividness in visual and olfactory imagery and higher visualization scores. Olfactory imagery may influence sociability. Moreover, a direct comparison between the two groups revealed that the sociable group had also higher verbalization scores than the unsociable group. It is worth noting that the social skill subscale exhibited relationships with imagery traits, almost entirely in the low-scoring group. In other words, these relationships are favorable for sociable individuals.

The social skill subscale of the AQ has been shown to predict several perceptual issues, including superior EFT performance [39,40] and reliance on high-spatial frequency features [33], which are understood as local processing biases. It also predicts impairment in abrupt object discrimination, which is attributed to reduced utilization of the dorsal stream to rapidly activate attention prior to ventral stream processing [41]. Further, it is indicative of a narrow temporal window and deficiency in multisensory integration [34]. It predicts longer evaluation times for unexpected words, which suggests a decreased availability of contextual information [42]. It is notable that all but the last example are perceptual issues, typified by a local processing bias, that cause weak social skills, whereas the last example is the outcome of weak social skills. Regarding imagery traits, the present study found that the AQ social skill subscale was a predictor for the low-scoring (sociable) group but not for the high-scoring (unsociable) group. Although social deficiency is a key symptom of ASD [1], the perceptual mechanisms that would cause social deficits may not be related to imagery qualities.

For the communication subscale, the high-scoring group (uncommunicative) had higher visual imagery vividness of a familiar shop scene but lower visualization and verbalization scores. Although the actual reason for the vividness of familiar shops is unknown, it appears that shops must be special for non-communicative individuals, making their images more vivid. Lower visualization and verbalization scores in an uncommunicative person should be understandable. Meanwhile, the low-scoring group (communicative) showed higher vividness of olfactory imagery. Olfactory imagery could play a role in "communicative" individuals, which is common among "sociable" individuals.

The communication subscale predicted several perceptual issues, including superior EFT performance [39] and reliance on high-spatial-frequency features [33], suggesting a local processing bias and a narrow temporal window [34]. Moreover, the communication subscale of the Autism Diagnostic Interview—Revised (ADI-R [43,44]) showed a negative correlation with an N200 amplitude in coherent motion processing, indicating dorsal stream deficiency [45]. These are unfavorable characteristics for uncommunicative individuals, which may be the basis for their lower visualization and verbalization.

For the attention-switching subscale, the high-scoring (unswitchable) group showed higher vividness of organic imagery in both comparative analyses. This finding suggests that weak attentional switching increases the vividness of organic imagery. Individuals with difficulties in switching attention may pay more attention to organic conditions or responses. The attention-switching subscale has been shown to predict reliance on high-spatial-frequency features, suggesting a local processing bias [33]. This bias may play an important role in weak attention-switching traits and increasing the vividness of organic imagery.

As cited above, it is notable that studies on the effectiveness of AQ subscales for the perceptual characteristics commonly observed, irrespective of varied subscales, a tendency typified by a detail-focused processing bias [33], including a narrow temporal window [34], an advantage in processing detail [39], a tendency to focus more on details [35], a local processing bias [33,34,39,40], and reduced utilization of the dorsal stream [41,45]. Our review here suggests that the "attention to detail" mechanism operates negatively for imagination, social skills, communications, and attention-switching, after the manner of AQ subscales, and would be a core basis of ASD. As for imagery qualities, our results revealed that a local processing bias makes visual images more vivid and makes imagination poorer.

We summarize the main findings regarding the associations between the AQ and imagery ability. The findings that exhibited contrasts between the high- and low-scoring groups of the AQ are as follows: (1) a higher degree of autistic-like traits as a whole is unfavorable for visualization, while a lower degree is favorable for it; (2) among the AQ subscales, a higher degree of unimaginativeness is unfavorable for imagery vividness across various modalities, visualization, and verbalization, while a lower degree is favorable for them; and (3) a higher degree of attention to detail is favorable for visual imagery vividness, while a lower degree is unfavorable for it. These are evidence that an individual's degree of autism-like traits affects their imagery traits.

The findings that did not exhibit contrasts between the high and low groups but did between either group and the other participants are as follows: for the high-scoring group, (1) a higher degree of being unable to switch attention is favorable for organic imagery vividness; and (2) a higher degree of uncommunicativeness is unfavorable for visualization. For the low-scoring group, (3) a lower degree of attention to detail is unfavorable for imagery vividness in visual, olfactory and whole modality, and visualization; (4) a lower degree of unsociableness (i.e., more sociable) is favorable for visual imagery and olfactory imagery vividness, as well as visualization and verbalization; and (5) a lower degree of uncommunicativeness (i.e., more communicative) is favorable for olfactory imagery vividness and verbalization. These indicate that the degree of autistic-like traits affects imagery traits in either of the end groups.

Table 7 exhibits a summary of the associations between the AQ and imagery ability. It reveals that a higher degree of autistic-like traits is unfavorable for imagery traits or is partly favorable, thereby elucidating how autistic-like traits influence imagery ability. On the other hand, Table 7 also reveals imagery characteristics in individuals with the lowest degree of autistic-like traits. The lowest degree is favorable for imagery traits or partly unfavorable. This shows, on the whole, opposite tendencies of individuals with a higher degree. Synthesizing the results of both degrees, the degree of autistic-like traits is a key factor in imagery ability. In other words, individual differences in autistic-like traits are closely related to individual differences in imagery ability.

Table 7. Summary of the associations between the AQ and imagery ability.

	High-Scoring Group			**Low-Scoring Group**		
	Vividness	**Verbalization**	**Visualization**	**Vividness**	**Verbalization**	**Visualization**
AQ total score			●			○
Social skill				○ (visual) ◦ (olfactory)	○	○
Attention switching	○ (organic)					
Attention to detail	○ (visual)			● (visual) • (olfactory) • (whole)		●
Communication			•	◦ (olfactory)	○	
Imagination	● (various modalities)	●	●	○ (various modalities)	○	○

(Note) • unfavorable for imagery ability; ○ favorable for imagery ability; small circle indicates marginal significance.

With respect to deficits in social communication and social interaction with ASD (DSM-5 [1]), it is worth noting that among the subscales of the AQ, imagination showed the most extensive relationships with imagery traits for both the high-scoring (unimaginative) and low-scoring (imaginative) groups, which contrasted sharply. As for two subscales which involve aspects of sociality, the communication subscale revealed lower visualization scores, though with marginal significance, in the high-scoring (uncommunicative) group, whereas social skill subscale, on the contrary, solely had relationships in the low-scoring (sociable) group. Concerning imagery qualities, these results suggest that imaginative deficits are quite involved in deficits in social communication and interaction in ASD. Symptom exemplifications such as deficits in non-verbal communicative behaviors and difficulties in imaginative play (DSM-5 [1]) are indicative of imagination deficits.

Concerning imagination, Crespi et al. [46] proposed the diametric disorders hypothesis that imagination is reduced in autism and increased in psychotic–affective conditions (schizophrenia, bipolar disorder, and depression). The study first discussed both these two sets of conditions in contrast for nine major aspects and correlates of imagination: (1) pretend play, (2) creativity and generativity, (3) narrative formation and comprehension, (4) mentalizing and empathizing, (5) meaning and salience, (6) episodic memory and future thinking, (7) mental imagery, (8) sensory abilities, and (9) neural activation and connectivity [46] (pp.183-190). Next, it examined the imagination subscale of the AQ and observed that nine of the ten items on this subscale provide a clear measure of brain default mode functions and involve aspects of social imagination [46] (p.190). Here, we can recognize the substances and correlates of imagination and recognize that imagination is specifically reduced in ASD, which the AQ subscale imagination measures optimally. At the same time, in that study, we notice that perceptual bases were referred to in many aspects and correlates of imagination, including reality-based creativity, e.g., [47]; attention focused on non-social aspects [48]; difficulties in switching attention [48]; increased attention to local details, e.g., [49–51]; increased focus on predictable patterns [52]; thinking in photorealistic pictures, e.g., [53]; rerunning images on a visuospatial scratch-pad [54]; the increased use of visuospatial processing strategies, e.g., [55]; and increased sensitivities and intense and accurate sensation [3]. Such perceptual bases of imagination in ASD resemble the cases, as discussed above, in which studies on the effectiveness of the AQ subscales for perceptual characteristics commonly observed detail-focused processing biases in all subscales. Baron-Cohen [56], in his systemizing theory of ASD, argued that sensory hypersensitivity is a basis of excellent attention to detail, which is in the service of strong systemizing. These all lead to the views that consider sensory features to be cores of ASD, as seen before [3–6,9–12,14,15]. Perceptual bases would be fundamental for imagination in ASD and, first of all, poverty of imagination in ASD would result from a detail-focused processing bias. Crespi et al. [46] appreciate Woodard and Van Reet's [57] psychological theory of autism based in imagination as highly compatible with previous theories. It proposes developing an imaginative ability which is determined by the degree of progression from part-object/inanimate object to whole-object/human object identification. As such, imagination would be the core phenotype of default mode functions.

According to Harari [58], the Cognitive Revolution occurred between 70,000 and 30,000 years ago, and *Homo sapiens* evolved imagination; that is, the cognitive capacity to believe in fiction. Matsuzawa [59] argued that the capacity to imagine differentiates humans from chimpanzees. Indeed, both a historian and a primate researcher have referred to imagination as a basic trait of humanity. However, individuals with ASD experience difficulties with this trait.

Regarding the distinctive sensory characteristics of ASD, which the DSM-5 [1] exemplifies hyper- or hyporeactivity to sensory input or an unusual interest in sensory aspects for ASD, the attention-to-detail subscale is worth exploring. Our study found that the high-scoring (attentive to detail) group showed higher vividness of visual imagery than the low-scoring (inattentive to detail) group. This finding suggests that a detail-focused processing bias makes visual images more vivid. Furthermore, the finding revealed that in another way, hyperactivity to sensory input creates vividness, whereas hyporeactivity does not. Arguably, in our results, the rather marginal but definite significance for vividness in attentive-to-detail individuals was compounded with more hyperactivity and less hyporeactivity in themselves. If that was the case, it would be essential to discuss the results with statistically marginal significance for the attentive-to-detail group with respect to imagery qualities.

This study provides several suggestions for investigating ASD. First, while research has tended to rely on the total AQ score, utilizing the AQ subscales may prove to be effective. This study derived significant findings and insights using these subscales. Second, to analyze the effects of AQ, this study recommends adopting the methods of comparing the high-scoring group with the rest of the participants (i.e., a randomly selected medium- and low-scoring combined group) and comparing the low-scoring group with the rest of the participants (i.e., a randomly selected high- and medium-scoring combined group). Thinning the participants is carried out to make the size of the control group similar to the target (experimental) group. A direct comparison between the high- and low-scoring groups, that is, the popular extreme group approach, has great problems [32]. Notably, the lower-end group may not necessarily be an appropriate control group because it excludes large number of individuals in the medium-scoring group on a continuum.

5. Limitations and Implications of this Study

Limitations regarding the present study should be addressed. First of all, this study used a sample of college students. Their cut-off point for the AQ total score, 26+, was approximately 20%, all of whom were in the high-scoring group, as seen in Table 2. But how our findings may be generalizable is unknown, which is a major limitation. In particular, it is not at all clear how this differs from a sample with more severe ASD traits.

Another is that the cut-off point for the total score, 26, roughly corresponds to the third quartile, so we used quartiles as criterion for classifying participants and applied that approach to the subscales. We assumed that it would be generally appropriate, but it may be unclear whether it is sufficient.

In terms of practical implications, this study may offer the following suggestion. Based on the finding that autistic-like traits are reflected in imagery traits, approaching individuals with ASD from a mental imagery perspective may be helpful. In other words, imagery tests may reveal the characteristics of the inner worlds of individuals with ASD.

6. Conclusions

This study is the first to explore relationships between the AQ and imagery tests. The findings indicate the following: (1) the imagery traits shown by the AQ total score are reflected in weak visualization and not in weak vividness. No noticeable differences in the vividness dimension would be attributable to setting off the effects of two traits, i.e., the merit of attention to detail against the demerit of imagination. (2) The imagination subscale shows, most extensively, negative relationships with imagery traits, whereas the communication subscale reveals a negative marginal relationship for visualization; the social skill subscale, on the contrary, is a predictor solely for the low-scoring (sociable) group. Such results suggest that imaginative deficits, in relation to imagery characteristics, would be quite involved in the deficits in social communication and interaction of ASD. (3) The degree of attention to detail primarily affects visual imagery vividness. The rather marginal but definite significance for the high-scoring group would be a compound with more hyperactivity and less hyporeactivity. (4) Given that detail-focused processing bias is a core basis of ASD, the bias makes visual images more vivid and imagination poorer. (5) The results of the lowest degree of autistic-like traits show, on the whole, opposite imagery tendencies to the higher degree. The results indicate that autistic-like traits are associated with qualities of the imagination and especially mental imagery ability.

Funding: This research received no external funding.

Institutional Review Board Statement: The study was conducted in accordance with the Declaration of Helsinki and approved by the Research Ethics Committee of Tohoku Bunkyo College (TB2023-10).

Informed Consent Statement: Informed consent was obtained from all participants involved in the study.

Data Availability Statement: The raw data supporting the conclusions of this article will be made available by the author upon request.

Acknowledgments: The author would like to acknowledge David F. Marks for help in preparing the manuscript.

Conflicts of Interest: The author declares no conflict of interest.

References

1. American Psychiatric Association. *DSM-5 Diagnostic and Statistical Manual of Mental Disorders*, 5th ed.; American Psychiatric Association: Washington, DC, USA, 2013.
2. Scott, F.J.; Baron-Cohen, S. Imagining real and unreal things: Evidence of a dissociation in autism. *J. Cogn. Neurosci.* **1996**, *8*, 371–382. [CrossRef] [PubMed]
3. Tavassoli, T.; Miller, L.; Schoen, S.A.; Nielsen, D.M.; Baron-Cohen, S. Sensory over-responsivity in adults with autism spectrum conditions. *Autism* **2014**, *18*, 428–432. [CrossRef] [PubMed]
4. Kumagai, T. *Autism and Hypersensitivity*; Shin-yo-sha: Tokyo, Japan, 2017. (In Japanese)
5. Takahashi, S.; Masubuchi, M. A study of real conditions and support of "hyper-sensitivity and insensibility" of persons with Asperger syndrome and high-functioning autism: Needs survey of persons with Asperger syndrome and high-functioning autism. *Bull. Tokyo Gakugei Univ. Edu. Sci.* **2008**, *59*, 287–310, (In Japanese with English abstract)
6. Crane, L.; Goddard, L.; Pring, L. Sensory processing in adults with autism spectrum disorders. *Autism* **2009**, *13*, 215–228. [CrossRef] [PubMed]
7. Brown, C.; Dunn, W. *Adult/Adolescent Sensory Profile: User's Manual*; Psychological Corporation: San Antonio, TX, USA, 2002.
8. Baron-Cohen, S.; Wheelwright, S.; Skinner, R.; Martin, J.; Clubley, E. The Autism-Spectrum Quotient (AQ): Evidence from Asperger syndrome/high-functioning autism, males and females, scientists and mathematicians. *J. Autism Dev. Disord.* **2001**, *31*, 5–17. [CrossRef] [PubMed]
9. Mayer, J.L. The relationship between autistic traits and atypical sensory functioning in neurotypical and ASD adults: A spectrum approach. *J. Autism Dev. Disord.* **2017**, *47*, 316–327. [CrossRef] [PubMed]
10. Robertson, A.E.; Simmons, D.R. The relationship between sensory sensitivity and autistic traits in the general population. *J. Autism Dev. Disord.* **2013**, *43*, 775–784. [CrossRef]
11. Takayama, Y.; Hashimoto, R.; Tani, M.; Kanai, C.; Yamada, T.; Watanabe, H.; Ono, T.; Kato, N.; Iwanami, A. Standardization of the Japanese version of the Glasgow Sensory Questionnaire (GSQ). *Res. Autism Spectr. Disord.* **2014**, *8*, 347–353. [CrossRef]
12. Sapey-Triomphe, L.-A.; Moulin, A.; Sonié, S.; Schmitz, C. The Glasgow Sensory Questionnaire: Validation of a French language version and refinement of sensory profiles of people with high Autism-Spectrum Quotient. *J. Autism Dev. Disord.* **2018**, *48*, 1549–1565. [CrossRef]
13. Schoen, S.A.; Miller, L.J.; Green, K.E. Pilot study of the Sensory Over-Responsivity Scales: Assessment and inventory. *Am. J. Occupa. Ther.* **2008**, *62*, 393–406. [CrossRef]
14. Tavassoli, T.; Hoekstra, R.A.; Baron-Cohen, S. The Sensory Perception Quotient (PSQ): Development and Validation of a New Sensory Questionnaire for Adults with and without Autism. *Mol. Autism* **2014**, *5*, 29. Available online: http://www.molecularautism.com/content/5/1/29 (accessed on 25 May 2021). [CrossRef]
15. Ujie, Y.; Wakabayashi, A. Psychometric properties and overlap of the GSQ and AQ among Japanese university students. *Int. J. Psychol. Stud.* **2015**, *7*, 195–205. [CrossRef]
16. Kosslyn, S.M.; Thompson, W.L.; Ganis, G. *The Case for Mental Imagery*; Oxford University Press: New York, NY, USA, 2006.
17. Wakabayashi, A.; Tojo, Y.; Baron-Cohen, S.; Wheelwright, S. The Autism-Spectrum Quotient (AQ) Japanese version: Evidence from high-functioning clinical group and normal adults. *Shinrigaku Kenkyu Jpn. J. Psychol.* **2004**, *75*, 78–84, (In Japanese with English abstract) [CrossRef]
18. Hatakeyama, T. Mechanisms of imagery ability: Identification on the basis of predictive effects of subjective imagery tests. *Jap. J. Ment. Imag.* **2019**, *17*, 13–36, (In Japanese with English abstract)
19. Galton, F. *Inquiries into Human Faculty and Its Development*; Macmillan: London, UK, 1883.
20. Betts, G.H. *The Distribution and Functions of Mental Imagery*; Teachers College, Columbia University: New York, NY, USA, 1909.
21. Sheehan, P.W. A shortened form of Betts Questionnaire upon Mental Imagery. *J. Clin. Psychol.* **1967**, *23*, 386–389. [CrossRef] [PubMed]
22. Marks, D.F. Visual imagery differences in the recall of pictures. *Br. J. Psychol.* **1973**, *64*, 17–24. [CrossRef] [PubMed]

23. Marks, D.F. Individual differences in the vividness of visual imagery and their effect on function. In *The Function and Nature of Imagery*; Sheehan, P.W., Ed.; Academic Press: New York, NY, USA, 1972; pp. 83–108.
24. Gordon, R. An investigation into some of the factors that favour the formation of stereotyped images. *Br. J. Psychol.* **1949**, *39*, 156–167. [CrossRef] [PubMed]
25. Paivio, A. *Imagery and Verbal Processes*; Holt, Rinehart and Winston: New York, NY, USA, 1971.
26. Richardson, A. Verbalizer-visualizer: A cognitive style dimension. *J. Ment. Imag.* **1977**, *1*, 109–126.
27. Tellegen, A.; Atkinson, G. Openness to absorbing and self-altering experiences ("absorption"), a trait related to hypnotic susceptibility. *J. Abnorm. Psychol.* **1974**, *83*, 268–277. [CrossRef] [PubMed]
28. Richardson, A. *Mental Imagery*; Routledge and Kegan Paul: London, UK, 1969.
29. Hatakeyama, T. Adults and children with high imagery show more pronounced perceptual priming effect. *Percept. Mot. Ski.* **1997**, *84*, 1315–1329. [CrossRef]
30. Kurita, H.; Koyama, T.; Osada, H. Autism-Spectrum Quotient-Japanese version a nd its short forms for screening normally intelligent persons with pervasive developmental disorders. *Psychiatry Clin. Neurosci.* **2005**, *59*, 490–496. [CrossRef] [PubMed]
31. Woodbury-Smith, M.R.; Robinson, J.; Wheelwright, S.; Baron-Cohen, S. Screening adults for Asperger syndrome using the AQ: A preliminary study of its diagnostic validity in clinical practice. *J. Autism Dev. Disord.* **2005**, *35*, 331–335. [CrossRef] [PubMed]
32. Preacher, K.J.; Rucker, D.D.; MacCallum, R.C.; Nicewander, W.A. Use of the extreme approach: A critical reexamination and new recommendations. *Psychol. Methods* **2005**, *10*, 178–192. [CrossRef]
33. Alink, A.; Charest, I. Clinically relevant autistic traits predict greater reliance on detail for image recognition. *Sci. Rep.* **2020**, *10*, 14239. [CrossRef] [PubMed]
34. Kawakami, S.; Uono, S.; Otsuka, S.; Zhao, S.; Toichi, M. Everything has its time: Narrow temporal windows are associated with high levels of autistic traits via weaknesses in multisensory integration. *J. Autism Dev. Disord.* **2020**, *50*, 1561–1571. [CrossRef] [PubMed]
35. Lassalle, A.; Itier, R.J. Autistic traits influence gaze-oriented attention to happy but not fearful faces. *Soc. Neurosci.* **2015**, *10*, 70–88. [CrossRef]
36. Marks, D.F. Phenomenological studies of visual mental imagery: A review and synthesis of historical datasets. *Vision* **2023**, *7*, 67. [CrossRef]
37. Richardson, A.; Patterson, Y. An evaluation of three procedures for increasing imagery vividness. In *International Review of Mental Imagery*; Sheikh, A.A., Ed.; Human Sciences Press: New York, NY, USA, 1986; Volume 2, pp. 166–191.
38. Hishitani, S. Imagery differences: What controls the vividness of imagery. In *Advances in Japanese Cognitive Science*; Japanese Cognitive Science Society, Ed.; Kodansha: Tokyo, Japan, 1993; Volume 6, pp. 81–117. (In Japanese with English abstract)
39. Grinter, E.J.; Van Beek, P.L.; Maybery, M.T.; Badcock, D.R. Brief report: Visuospatial analysis and self-rated autistic-like traits. *J. Autism Dev. Disord.* **2009**, *39*, 670–677. [CrossRef]
40. Russell-Smith, S.N.; Maybery, M.T.; Bayliss, D.M.; Sng, A.A.H. Support for a link between the local processing bias and social deficits in autism: An investigation of Embedded Figure Test performance in non-clinical individuals. *J. Autism Dev. Disord.* **2012**, *42*, 2420–2430. [CrossRef]
41. Laycock, R.; Cross, A.J.; Nogare, F.D.; Crewther, S.G. Self-rated social skills predict visual perception: Impairments in object discrimination requiring transient attention associated with high autistic tendency. *Autism Res.* **2014**, *7*, 104–111. [CrossRef] [PubMed]
42. Ishikawa, M.; Itakura, S.; Tanabe, H.C. Autistic traits affect P300 response to unexpected events, regardless of mental state inferences. *Autism Res. Treat.* **2017**, *2017*, 8195129. [CrossRef]
43. Boelte, S.; Ruehl, D.; Schmoetzer, G.; Poustka, F. *ADI-R Diagnostisches Interview für Autism-Revidiert. Deutsche Fassung des Autism Diagnostic Interview-Revised*; Hogrefe: Göttingen, Germany, 2006.
44. LeCouteur, A.; Rutter, M.; Lord, C.; Rios, P.; Robertson, S.; Holdgrafer, M.; McLennan, J. Autism Diagnostic Interview: A standardized investigator-based instrument. *J. Autism Dev. Disord.* **1989**, *19*, 363–387. [CrossRef]
45. Greimel, E.; Bartling, J.; Dunkel, J.; Brückl, M.; Deimel, W.; Remschmidt, H.; Kamp-Becker, I.; Schulte-Körne, G. The temporal dynamics of coherent motion processing in autism spectrum disorder: Evidence for a deficit in the dorsal pathway. *Behav. Brain Res.* **2013**, *251*, 168–175. [CrossRef]
46. Crespi, B.; Leach, E.; Dinsdale, N.; Mokkonen, M.; Hurd, P. Imagination in human social cognition, autism, and psycho-affective conditions. *Cognition* **2016**, *150*, 181–199. [CrossRef]
47. Craig, J.; Baron-Cohen, S. Creativity and imagination in autism and Asperger syndrome. *J. Autism Dev. Disord.* **1999**, *29*, 319–326. [CrossRef] [PubMed]
48. Ploog, B.O. Stimulus overselectivity four decades later: A review of the literature and its implications for current research in autism spectrum disorder. *J. Autism Dev. Disord.* **2010**, *40*, 1332–1349. [CrossRef]
49. Barnes, J.L.; Baron-Cohen, S. The big picture: Story-telling ability in adults with autism spectrum conditions. *J. Autism Dev. Disord.* **2012**, *42*, 1557–1565. [CrossRef]
50. Happé, F.; Frith, U. The weak coherence account: Detail-focused cognitive style in autism spectrum disorders. *J. Autism Dev. Disord.* **2006**, *36*, 5–25. [CrossRef]
51. Jolliffe, T.; Baron-Cohen, S. A test of central coherence theory: Linguistic processing in high-functioning adults with autism or Asperger syndrome: Is local coherence impaired? *Cognition* **1999**, *71*, 149–185. [CrossRef] [PubMed]

52. Baron-Cohen, S.; Lombardo, M.V.; Auyeung, B.; Ashwin, E.; Chakrabarti, B.; Knickmeyer, R. Why are autism spectrum conditions more prevalent in males? *PLoS Biol.* **2011**, *9*, e1001081. [CrossRef] [PubMed]
53. Grandin, T. How does visual thinking work in the mind of a person with autism? A personal account. *Philos. Trans. R. Soc. B Biol. Sci.* **2009**, *364*, 1437–1442. [CrossRef]
54. Boucher, J. Memory and generativity in very high functioning autism: A firsthand account, and an interpretation. *Autism* **2007**, *11*, 255–264. [CrossRef]
55. Sahyoun, C.P.; Belliveau, J.W.; Soulières, I.; Schwartz, S.; Mody, M. Neuroimaging of the functional and structural networks underlying visuospatial vs. linguistic reasoning in high-functioning autism. *Neuropsychol.* **2010**, *48*, 86–95. [CrossRef] [PubMed]
56. Baron-Cohen, S.; Ashwin, E.; Ashwin, C.; Tavassoli, T.; Chakrabarti, B. Talent in autism: Hyper-systemizing, hyper-attention to detail and sensory hypersensitivity. *Phil. Trans. R. Soc. B.* **2009**, *364*, 1377–1383. [CrossRef] [PubMed]
57. Woodard, C.R.; Van Reet, J. Object identification and imagination: An alternative to the meta-representational explanation of autism. *J. Autism Dev. Disord.* **2011**, *41*, 213–226. [CrossRef] [PubMed]
58. Harari, Y.N. *Sapiens: A Brief History of Humankind;* Deborah Harris Agency: Jerusalem, Israel, 2011.
59. Matsuzawa, T. *The Power of Imagining: Human Mind What Chimpanzees Taught Us;* Iwanami Shoten: Tokyo, Japan, 2011. (In Japanese)

vision

MDPI

Article

The Neural Basis of a Cognitive Function That Suppresses the Generation of Mental Imagery: Evidence from a Functional Magnetic Resonance Imaging Study

Hiroki Motoyama [1,*] and Shinsuke Hishitani [2]

1 College of Humanities and Social Sciences, Ibaraki University, Mito 3108512, Japan
2 Emeritus Professor of Psychology, Hokkaido University, Sapporo 0600810, Japan; hishitani@gol.com
* Correspondence: hiroki.motoyama.m@vc.ibaraki.ac.jp

Abstract: This study elucidated the brain regions associated with the perception-driven suppression of mental imagery generation by comparing brain activation in a picture observation condition with that in a positive imagery generation condition. The assumption was that mental imagery generation would be suppressed in the former condition but not in the latter. The results show significant activation of the left posterior cingulate gyrus (PCgG) in the former condition compared to in the latter condition. This finding is generally consistent with a previous study showing that the left PCgG suppresses mental imagery generation. Furthermore, correlational analyses showed a significant correlation between the activation of the left PCgG and participants' subjective richness ratings, which are a measure of the clarity of a presented picture. Increased activity in the PCgG makes it more difficult to generate mental imagery. As visual perceptual processing and visual imagery generation are in competition, the suppression of mental imagery generation leads to enhanced visual perceptual processing. In other words, the greater the suppression of mental imagery, the clearer the presented pictures are perceived. The significant correlation found is consistent with this idea. The current results and previous studies suggest that the left PCgG plays a role in suppressing the generation of mental imagery.

Keywords: mental imagery; visual perception; vividness; richness; suppression; posterior cingulate gyrus

1. Introduction

What processes are involved in the generation of mental imagery? Most imagery researchers agree that conscious mental imagery is represented in the image construction stage through transformations applied to information retrieved from long-term memory (LTM) [1–6]. Kosslyn (1980; 1994) referred to the image construction stage as the visual buffer, stating that images formed in that buffer on the basis of information from LTM are mental images, whereas images formed in that buffer on the basis of information from the outside world are visual perceptual images [2,7]. This definition suggests that the image construction stage is a common process within mental imagery and visual perceptual imagery processing. This claim has been supported by numerous studies showing a competitive relationship between the visual perceptual process and the generation of visual imagery [8–22]. This is because these two processes share a common process, such that carrying out one process disrupts the other, resulting in competition. If this is the case, then for us to act adaptively, we would need a mental function that suppresses one process by not forming mental images while visual perceptual images are actively being formed, or vice versa, such as not forming visual perceptual images while mental images are being formed. Kosslyn (1987) suggested that visual hallucinations are produced when memory inputs from LTM to the visual buffer are strong enough to overwhelm perceptual inputs. If this is correct, then a function that suppresses imagery generation so as not to produce visual hallucinations may be active as we construct the visual world in everyday life [23]. The current study investigates the neural basis that directly supports the existence of this perception-driven imagery suppression mechanism.

Motoyama and Hishitani (2016) demonstrated the imagery suppression mechanism in a research context that was slightly different from that used in the present study [24]. They conducted an experiment to locate brain regions that are associated with the suppression mechanism for negative imagery generation. Previous studies have reported that the vividness of mental imagery varies according to the emotional valence of the imagined object, such that negative imagery is more vague than positive imagery [24–26] [1]. Participants in the above studies were simply asked to intentionally generate positive or negative imagery, but they were not instructed to intentionally control the vividness of the imagery, regardless of whether the imagery generated was positive or negative. Although the participants did not intend to generate vague negative imagery or vivid positive imagery, negative imagery was more vague than positive imagery. Hishitani (1995) and Hishitani et al. (2011) argued that, when the emotional valence of an object or an event is negative, imagery generation processing is suppressed because generating vivid negative imagery is stressful. As a result, the mechanism that suppresses negative imagery generation is activated during negative imagery generation, and negative imagery becomes more vague than positive imagery [1,27] [2].

Motoyama and Hishitani (2016) attempted to identify the brain areas in which activation is significantly greater during negative imagery generation than during positive imagery generation, as they expected the suppression of mental imagery generation to be more active during negative imagery generation than during positive imagery generation [24]. They first showed that imagery generated in a positive condition was rated as more vivid than imagery generated in a negative condition. They then showed significantly increased activation in a part of the left posterior cingulate gyrus (PCgG) during the negative imagery generation condition compared to the positive imagery generation condition. In addition, the activity in this region was negatively correlated with the vividness of the negative imagery generated. That is, the greater the activity in this region, the less vivid the images in the negative imagery condition. If the left PCgG functions as a suppressor of the generation of mental imagery, especially negative imagery, then the results would be consistent with the following explanation: An increase in PCgG activity during negative imagery generation indicates a stronger suppressive function in the negative imagery condition. Therefore, the higher the activity of this brain area, the lower the vividness of negative imagery. These findings suggest that some areas of the left PCgG are involved in the emotion-driven suppression of imagery, which prevents the generation of imagery to avoid negative emotions. Furthermore, there was no correlation between PCgG activity and the vividness of positive imagery. This was interpreted as meaning that positive imagery was not necessary for emotional avoidance and that emotion-driven imagery suppression did not occur.

If the left PCgG is involved in imagery suppression in general, then it would be activated not only during the formation of negative imagery but also during other activities that require the suppression of the formation of mental imagery. That is, it is expected that at least some regions of the left PCgG would be activated while a person looks at visual perceptual images because of the competitive relationship between visual perceptual processing and visual imagery generation, as discussed above. Therefore, this study examines whether brain areas in the left PCgG are more activated in a visual object observation condition than in a positive imagery generation condition, in which mental imagery generation is considered not to cause, or mostly not to cause, the emotion-driven suppression of imagery. If there was a significantly activated area in the left PCgG, then it would provide evidence that this brain area is involved in the suppression of imagery generation during visual perceptual processing, i.e., the perception-driven suppression of imagery generation. It might be suggested that the resting period, during which there appears to be no mental activity, should be used as a control condition instead of the positive imagery generation condition. However, the rest period is not appropriate to be used as a control condition in this study. The PCgG is part of the default mode network that is activated at rest. At present, it is not possible to distinguish whether the activity observed in the left PCgG at rest reflects the automatic suppression of imagery generation or mental activity other than imagery suppression. The control condition in this study requires the following two conditions: (1) a picture is shown to provide the same perceptual information as in the picture observation condition but not to activate the perception-driven suppression of imagery, and (2) the emotion-driven suppression of imagery does not function. Therefore, it is considered an appropriate control condition to simply see a presented picture but not to perceptually process it and to generate positive imagery that is not the same as the presented picture.

In the picture observation condition in this study, a participant looks at a presented picture and evaluates its richness, i.e., the degree of clarity of the presented picture. Assuming that some brain areas in the left PCgG would be activated in the picture observation condition compared to in the control condition mentioned above, it could be speculated that the activated brain area plays a role in the perception-driven suppression of imagery. In this case, it is expected that there would be a correlation between the activity in the brain area and the participants' richness ratings, which represent the subjective vividness of the presented picture in the picture observation condition. Several previous studies have shown that visual imagery acts as a disturbance in the detection and perception of presented visual stimuli [1,8,21,28]. Therefore, if the activated brain area in the left PCgG is active, then imagery generation processing would be suppressed, and visual perceptual processing, which competes with mental imagery processing, would be facilitated. In other words, if the left PCgG is more active, then the presented visual object would be perceived as being clearer because less mental imagery would be generated. Conversely, if the activity in the left PCgG region is weaker, then the presented visual object would be perceived as being more vague. If the above correlation between brain area activity and participants' richness ratings is obtained, then it would be evidence that the activated brain area in the left PCgG plays a role in the perception-driven suppression of imagery generation.

Furthermore, if the activated brain area plays a role in imagery suppression, then some might think that there would be a negative correlation between the activation of the brain area and imagery vividness ratings. However, this is not the case in this experiment. This is because, firstly, the imagery generation condition in this study is set up as a condition in which the emotion-driven suppression caused by negative emotions is not, or almost not, produced. Secondly, because participants are not required to actively process the visual perceptual stimuli of the outside world, the perception-driven suppression of imagery will not occur during positive imagery generation. Therefore, there will be no correlation between the activation of the relevant brain area and the vividness of the imagery.

This study tested the following two hypotheses: (1) compared to the positive imagery condition, in which participants see a picture and generate imagery unrelated to it, stronger activity in the left PCgG is observed when participants observe visual perceptual information in the outside world without generating imagery (picture observation condition), and (2) there is a correlation between activity in the left PCgG and richness ratings. In addition, this study tested the hypothesis stating that there is no correlation between the vividness of positive imagery and left PCgG activity.

2. Materials and Methods

2.1. Participants

Healthy right-handed volunteers (n = 16, 5 females and 11 males with an average age of 26.8 years, SD = 5.7, ranging from 21 to 45 years) participated in this study, and each was paid 2000 JPY ($20) for their participation. The participants were undergraduate or graduate students at Hokkaido University. The Ethics Committee of Hokkaido University, Graduate School of Letters, approved the experimental protocol of this study. In addition, all participants gave written informed consent before participating. The handedness of the participants was assessed using the Japanese translation of the Edinburgh Handedness Survey [29], which indicated that all the participants were right-handed.

2.2. Materials

The words used to generate imagery in the positive imagery generation condition (Appendix A), which were not expected to activate the emotion-driven suppression of imagery function, were positive words that were used in [24]. The emotional valence of these words was based on the norms in [30], in which the emotional valence of a noun is rated on a 7-point Likert-type scale (where 1 indicates extremely negative emotions, and 7 indicates extremely positive emotions). The mean emotional valence of the words used in the positive imagery generation condition was 5.12 (ranging from 4.63 to 5.67), which was significantly positive compared to the neutral value, corresponding to a score of 4 ($t(15) = 14.704$, $p < 0.00001$). To prepare the pictures (Appendix B) presented in the positive imagery generation condition, nouns were selected from neutral words with an approximate neutral value of 4 (the mean emotional valence was 4.09, ranging from 3.92 to 4.24) in [30].

In the visual perceptual processing condition, a picture was shown on a screen that the participants were asked to look at (picture observation condition). The pictures presented corresponded to the words (Appendix A) used in the positive imagery generation condition. In the positive imagery generation condition, a noun appeared on the screen, and the participants were asked to form a mental image of the noun's referent while seeing a picture. The pictorial stimuli (Appendices A and B) that were presented in both conditions were selected from ART EXPLOSION 750,000 Macintosh edition [31]. A Gaussian filter was applied to the prepared pictures at 4 levels (Gaussian effect 0 mm radius: no filtering, 4 mm, 8 mm, and 12 mm) by using Adobe Photoshop CS4, and the clarity of the presented pictures was varied. The higher the Gaussian effect, the greater the effect of the low-pass filter and the lower the clarity of the picture.

2.3. Apparatus

SuperLab Pro 4.4 (Cedrus Corporation, San Pedro, CA, USA) running on an Apple MacBook computer was used to present the stimuli and record the participants' responses. An MRI-compatible response box [32] was used to record the participants' subjective ratings of the vividness of the mental imagery and the richness of the pictures.

2.4. Procedure

The procedure used in this study included two activation conditions and a baseline condition (Figure 1). In the picture observation condition, the participants actively observed visual perceptual information in the external world without generating any imagery, whereas in the positive imagery generation condition, they performed a mental imagery generation task while seeing a picture. In this condition, the participants were asked to generate imagery while ignoring the presented picture. In the baseline condition, the participants were asked to relax without generating any imagery or looking at pictures.

If the participants' task in the positive imagery generation condition had been to generate positive imagery without the input of any visual perceptual information from the outside world, then the brain activity during positive imagery generation would have been measured without the participants seeing a picture, whereas that during picture observation would have been measured with the participants looking at a picture. Therefore, not only would the presence or absence of the activity of the perception-driven imagery suppression function differ between the two conditions but the presence or absence of the input visual perceptual information would also differ. In this case, even if there was a difference in PCgG activity between the two conditions, it would not be clear whether it reflected the activity of the perception-driven suppression of imagery or the processing of visual perceptual information. Therefore, visual perceptual information was input in both the positive imagery generation and picture observation conditions, and the participants were asked to generate imagery without actively observing it in the former condition and to actively observe it in the latter condition.

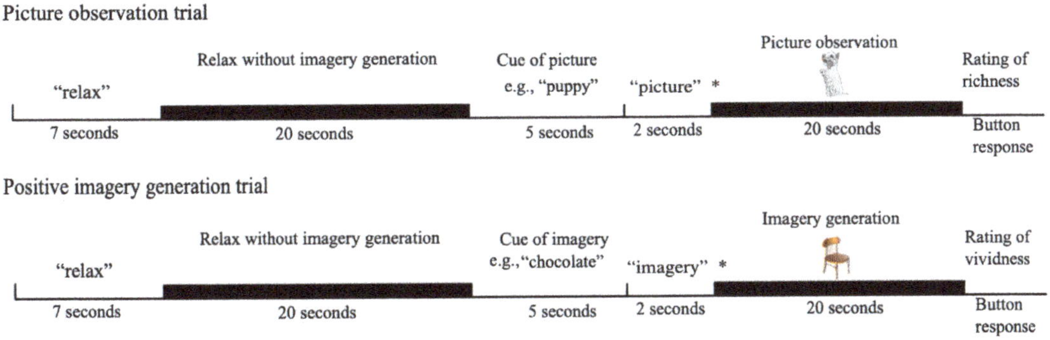

Figure 1. The experimental procedure: The asterisk (*) represents the cue for the participant's task for the next 20 s. "Picture" indicates that their task was to observe the presented picture. "Imagery" indicates that their task was to generate imagery of the cue while seeing at the presented picture. The pictures presented in the picture observation and positive imagery generation conditions were, for example, a puppy and a chair.

The participants viewed the screen through a mirror system that was mounted onto a head coil, and their eyes were open while the experiment was conducted. In each picture observation trial, the word "relax" appeared on the screen for seven seconds, followed by a grey blank screen with no visual stimuli that was presented for twenty seconds. The participants were instructed to relax during the presentation of the blank screen (baseline condition). Then, a noun appeared on the screen for five seconds, followed by the word "picture", which appeared on the screen for two seconds, and then a picture of the previously presented noun, which appeared for twenty seconds. The participants were asked to look at the picture when the word "picture" was presented after the noun. A message then appeared on the screen asking the participants to rate the richness of the picture. The participants rated the richness of the presented picture on a 4-point scale ranging from 1 (*very vivid*) to 4 (*very vague*)[3]. The responses were recorded on a computer using a response box placed next to the participants' left hand. The next trial started when the participants pressed any button on the response box. The positive imagery generation trials were identical to the picture observation trials until the noun was presented after the baseline condition. In the positive imagery generation trials, the word "imagery" appeared on the screen for two seconds after the noun disappeared, and an image unrelated to the previously presented noun appeared for 20 s. The participants were asked to generate imagery of the referent of the noun. They were also asked to experience the positive emotion evoked by the generated imagery in order to reduce the emotion-driven suppression of imagery function. On this occasion, the participants were asked to generate imagery by superimposing imagery on the presented picture rather than by making associations with it. For example, if the participants generated an image of a kitten while seeing a picture of a bus, they were asked not to form an image of a kitten sitting on the roof of the bus or in the bus but rather to form an image in which the kitten was superimposed on the picture of the bus so that the picture of the bus formed the background of the image. In this condition, the participants were asked to ignore the incoming visual perceptual information while the picture was presented. However, it was possible for the participants to look elsewhere on the screen, where the picture was not presented, rather than at the presented picture, when they were asked to overtly ignore the presented picture. In this case, no visual information was provided to the participants. Thus, the participants were given the above instruction. This instruction was given to facilitate a common cognitive activity between the two conditions of picture observation and positive imagery generation. Then, a screen appeared asking about the vividness of the imagery, and the participants were asked to rate the vividness of their mental imagery on a 4-point scale ranging from 1 (*very vivid*) to 4 (*very vague*). A diagram of the present study is shown in Figure 1. Each session consisted of eight trials (four picture observation trials and four positive imagery generation trials). The entire experiment consisted of two sessions. The order of the picture observation and positive imagery generation trials was randomised. The words and pictures used in the picture observation and positive imagery generation conditions were also randomised for each participant. The explanation of the experimental procedure was repeated until the participants thoroughly understood the task, and practice trials were conducted before the experimental trials.

2.5. *fMRI Data Acquisition and Data Analysis*

2.5.1. fMRI Data Acquisition

A whole-body 1.5 T Signa Echo-Speed scanner (General Electric, Inc. Chicago, IL, USA) was used to acquire high-resolution spin echo T1-weighted anatomical images (20 axial slices, matrix size = 256 × 256, TR = 500 ms, FOV = 240 × 240, voxel size = 0.93 mm isotropic) and gradient echo echo-planar T2*-weighted images with a BOLD contrast (20 axial slices, matrix size = 64 × 64, flip angle = 90°, TR = 3000 ms, TE = 40 ms, FOV = 240 × 240 mm, voxel size = 3.75 mm isotopic, slice thickness = 4 mm, and slice gap = 0.8 mm). A total of 314 scans were acquired per participant (157 volumes × 2 sessions). Brain activity was measured throughout the session.

2.5.2. fMRI Data Analysis

Data were analysed via statistical parametric mapping (SPM8, www.fil.ion.ucl.ac.uk/spm/software/spm8, accessed on 8 January 2024, MATLAB version R2011b) using a general linear model. All functional volumes were realigned to each participant's first volume to correct for head motion, which was spatially normalised to the Montreal Neurological Institute (MNI) brain template. Each voxel was resampled to 2 × 2 × 2 mm and smoothed with an 8 mm full-width-at-half-maximum Gaussian kernel. Functional data were analysed using a block design. For a group analysis, a random effects analysis was performed based on the general linear model. Each block was defined as a regressor of interest: picture observation, positive imagery generation, relax. A high-pass filter with a cut-off period of 128 s was used to remove low-frequency noise. We predefined the left PCgG as a region of interest (ROI) and prepared templates using the WFU PickAtlas tool 3.0.5 [33,34] according to an Automatic Anatomical Labeling (AAL) atlas previously reported in [35] (Figure 2). And we performed small volume correction (SVC) analyses on the ROI. We investigated whether stronger activity occurred in the left PCgG during picture observing (20 s) in the picture observation condition compared to imagery generation while seeing a picture (20 s) in the positive imagery generation condition. Brain activity during picture observation and imagery generation was averaged for each condition and participant. Brain activity during picture observation in the picture observation condition (or during positive imagery generation in the positive imagery generation condition) was the average activity of each participant over 8 trials (2 trials for each Gaussian filter of 0, 4, 8, and 12). As described above, although the main aim of this experiment was to compare brain activity during picture observation − during positive imagery generation, we also investigated whether stronger activities occurred in the left PCgG during positive imagery generation compared to during relaxation (20 s) in the baseline condition and during picture observation compared to during relaxation (20 s) in the baseline condition. In the present study, it was assumed that the left PCgG would be more active in a condition in which the suppression of imagery generation works more strongly (i.e., the picture observation condition) than in a condition in which it does not work (or only weakly works) (i.e., the positive imagery generation condition). However, it is also possible that the brain activity that occurs in such conditions reflects cognitive activities other than the suppression of imagery generation.

For example, it may reflect visual perceptual and emotional processing, which may occur in the picture observation and positive imagery generation conditions. If this is the case, then stronger left PCgG activity should also occur in the positive imagery generation and picture observation conditions than in the baseline condition, in which positive emotion and visual perceptual processing would not occur. Therefore, we decided to compare brain activity during positive imagery generation − during relaxation in the baseline condition and during picture observation − during relaxation in the baseline condition. The baseline condition that was compared with the activation conditions (picture observation and positive imagery generation) was during relaxation in the baseline condition immediately before each activation condition. Statistical significance was set at $p < 0.05$, and the family-wise error (FWE) was corrected at the voxel level within small volume correction. In addition, the time course of the MRI signal was calculated using Marsbar [36].

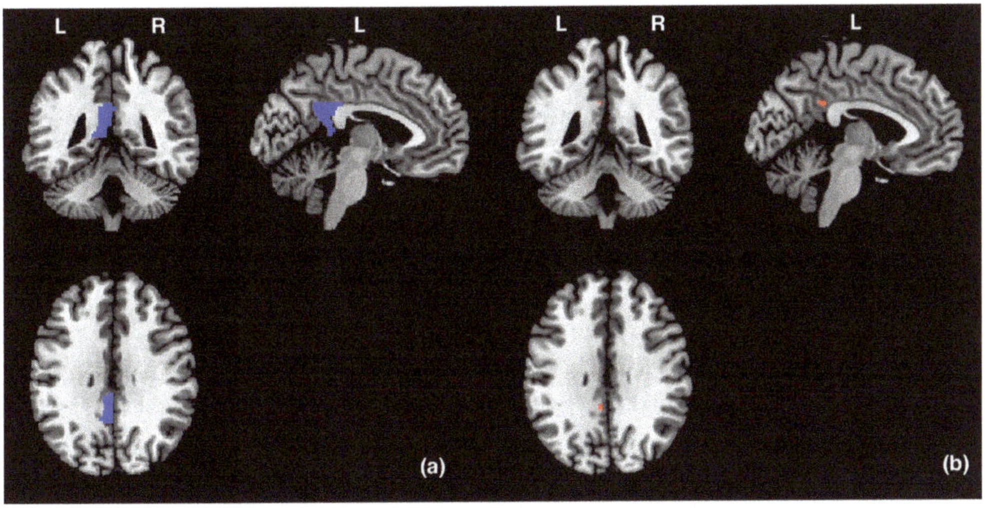

Figure 2. (**a**) The blue areas represent the ROI (the left PCgG) in this study, and the size was 3704 mm^3 (463 voxels). (**b**) The red areas represent the areas (−4, −44, 32) in the left posterior cingulate gyrus that were significantly activated during picture observation compared to during positive imagery generation. L and R indicate left and right hemispheres, respectively.

3. Results

3.1. Brain Activity in Left Posterior Cingulate Gyrus

In order to identify the specific brain area involved in the suppression of imagery generation, we measured the brain activity in the left PCgG, whose activity was expected to be significantly greater during picture observation in the picture observation condition than during imagery generation in the positive imagery generation condition. As shown in Table 1 and Figure 2, significant activity (−4, −44, 32) was found (p (FWE) = 0.016).

Table 1. Areas of significant activation observed during picture observation compared to during positive imagery generation.

Brain Area Brodman Area (BA)	Number of Voxels in Cluster	Voxel−Level p Value FWE−Corrected	Voxel−Level p Value Uncorrected	T Value at Local Maximum	MNI Coordinates x	y	z
Posterior Cingulate Gyrus (BA31)	4	0.016	0	4.71	−4	−44	32

In addition, brain activity was measured in the left PCgG during positive imagery generation – during relaxation in the baseline condition, and during picture observation – during relaxation in the baseline condition. No significant activity was found in either comparison.

3.2. Relationship between MRI Signal Changes in the Activated Area (−4, −44, 32) and the Participants' Subjective Richness Ratings of the Pictures

If the activated brain area in the left PCgG plays a role in the perception-driven suppression of imagery generation, then there would be a correlation between the richness ratings, which are subjective assessments of the clarity of the presented pictures, and the activation levels in the activated brain area, such that increased activity in the activated brain area would be associated with a higher level of richness. Therefore, the correlation between the activity in the activated brain area and the richness ratings was examined. Each participant completed eight trials in the picture observation condition, and eight richness ratings were obtained for each participant, from which the mean richness rating for each participant was calculated. We calculated the correlation coefficient between the mean richness ratings and brain activity during picture observation for each of the 16 participants.

We calculated the mean changes in the MRI signal as follows: First, we calculated the MRI signals in the brain area examined in this study using Marsbar. Then, we calculated the changes in the MRI signal during picture observation relative to those during relaxation. Relaxation in the baseline condition lasted for 20 s before the picture observation, which also lasted for 20 s. We calculated changes in the MRI signal during picture observation in the picture observation condition relative to those during relaxation in the most recent baseline condition. In this experiment, the TR duration was 3 s. Thus, the MRI signal data were recorded 6–7 times each during relaxation and during picture observation. The data from only six scans from the beginning to the end of an activity were analysed. These six scans were averaged for each activity. The MRI signal change was then calculated for the averaged MRI signal during picture observation relative to that during relaxation. The mean percentage change in the MRI signal and the mean richness score for the picture observation condition were calculated for each participant. That is, the following means were calculated for each participant across the eight picture observation trials: percentage change in the MRI signal and richness score. Shapiro–Wilk tests were performed on the mean percentage change in the MRI signal and the mean richness scores to test for normality. As a result, Spearman's rank correlation was calculated, as the mean richness scores were not normally distributed ($p = 0.06$). There was a significant negative correlation between the mean percentage change in the MRI signal and the mean richness score (r_s (14) = −0.50, $p = 0.047$, two tails, Figure 3). This result indicates that the stronger the activation of the brain area in the left PCgG, the clearer the recognition of the presented picture.

3.3. Relationship between MRI Signal Changes in the Activated Area (−4, −44, 32) and the Vividness of Imagery

Although a previous study suggests that there is no such correlation, if the activated brain area in the left PCgG were to play a role in suppressing the generation of any type of mental imagery, then there would be a correlation between the vividness of the imagery and the activity in the brain area, such that increased activity in the brain area would be associated with a reduced vividness of the positive imagery generated by the participants in this experiment. Therefore, the correlation coefficient between the MRI signal changes in the cluster containing (−4, −44, 32) and the subjective vividness of the imagery was calculated. The mean change in the MRI signal was calculated during positive imagery generation relative to that during relaxation in the most recent baseline condition. The method used to calculate the mean percentage signal change was identical to that described in Section 3.2. The mean subjective rating of the imagery vividness was calculated for each participant as follows: Each participant took part in eight positive imagery generation trials, and eight imagery vividness ratings were obtained for each participant. The mean of the imagery vividness ratings from these eight trials was taken as the mean imagery vividness rating for each participant. As there were 16 participants in this study, the mean imagery vividness ratings and brain activity during positive imagery generation for each participant resulted in 16 data points. The correlation coefficient was calculated between the two variables. The mean percent signal change and mean imagery vividness ratings were calculated for each participant, and a correlation analysis was performed on the two variables. A Pearson's correlation coefficient was calculated because the Shapiro–Wilk tests conducted on the mean percentage change in the MRI signal and the mean imagery vividness ratings did not reject the null hypothesis, indicating that both variables were normally distributed. The results indicate that there is no significant correlation between the variables (r (14) = −0.07, $p > 0.10$).

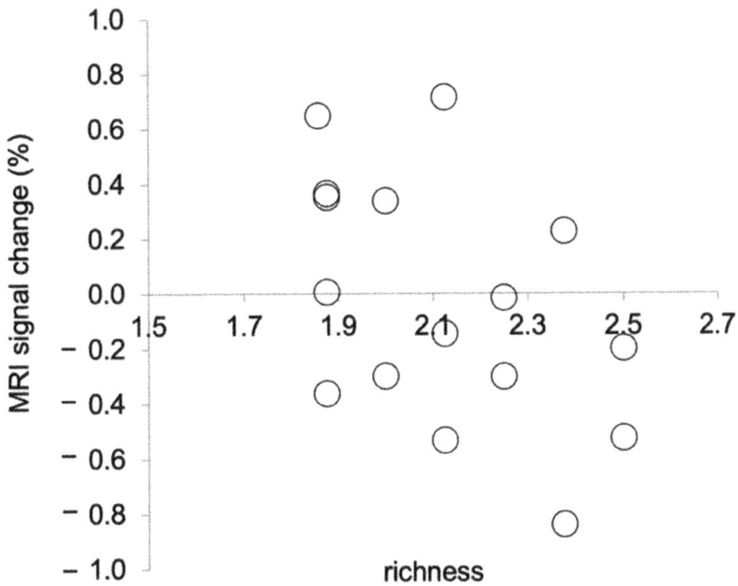

Figure 3. Correlation between MRI signal changes at (−4, −44, 32) during picture observation and richness ratings. A lower richness rating means that the presented picture was perceived as clearer.

4. Discussion

This study attempted to identify the brain regions involved in the perception-driven suppression of imagery generation. A previous study suggested that the left PCgG is involved in the suppression of negative imagery generation. If the left PCgG were involved in the suppression of imagery in general, it would be activated not only during the formation of negative imagery but also during other activities that require the suppression of the formation of mental imagery. It has been shown that the processes of visual perception and visual imagery formation are in a competitive relationship. That is, it is expected that at least some regions of the left PCgG would be activated during the observation of visual perceptual stimuli because the formation of mental imagery would be suppressed while visual perceptual images are actively observed.

In this study, part of the left PCgG was more active during the picture observation condition, during which the suppression of perception-driven imagery generation was considered to be stronger than during the positive imagery generation condition. Furthermore, there was a significant negative correlation between the activity in the left PCgG and the perceptual richness ratings. More specifically, the higher the activity in the brain area, the lower the richness ratings (i.e., the clearer the picture). Additionally, the lower the activity in the brain area, the higher the richness ratings. If we assume that the left PCgG plays a role in suppressing the generation of mental imagery, then we can offer a consistent explanation for these results. That is, when the brain area concerned was more strongly activated and the suppression of imagery generation function was working more strongly, visual perceptual processing was more activated because competitive mental imagery generation did not occur. Therefore, a visual object was perceived as clearer when the relevant brain area was more strongly activated. These results suggest that the left PCgG plays a role in the suppression of imagery generation.

Motoyama and Hishitani (2016) previously demonstrated the imagery suppression mechanism in a research context slightly different to that used in the present study [24]. They showed significantly increased activation in a part of the left PCgG during negative imagery generation compared to during positive imagery generation. Their finding indicates that the activity in the left PCgG increased during negative imagery generation when the emotion-driven suppression of imagery generation was stronger. From a perspective different to that of Motoyama and Hishitani (2016) [24], the result of the current study shows that the left PCgG plays a role in the suppression of imagery generation. However, the activated brain coordinates found in this study do not perfectly match the area involved in the suppression of mental imagery generation shown in [24]. The peak Talairach coordinates of the activated brain area in [24] were (−18, −53, 30), and the difference in the peak coordinates between this study and those in [24] is about 18.5 mm, suggesting that the perception-driven and emotion-driven suppression of imagery generation may be caused by different mechanisms. Since this possibility is only based on the results of [24] and the present study, it needs to be examined in further experiments.

The present study found that the left PCgG was more strongly activated in the picture observation condition than in the positive imagery condition. This finding provides evidence for the notion that the left PCgG plays a role in the perception-driven suppression of imagery generation. However, if an increase in visual perceptual processing could lead to an increase in PCgG activity, then the result could reflect the visual perceptual information processing activity that occurs in the PCgG. This interpretation can be ruled out for two reasons. First, if the activity observed in the left PCgG in this experiment reflected visual perceptual processing, then the activity in this brain area would also have been produced in the picture observation condition—in the baseline condition. However, no such brain activity was found. Second, a previous study suggested the possibility that the PCgG does not play a role in the processing of visual perceptual information, such as in the observation of a picture in this experiment. It has been postulated that PCgG activity is reduced when paying attention to the outside world [37–39], suggesting that PCgG activity is reduced when actively looking at pictures, as in the task used in this experiment. In addition, it has been suggested that similar levels of activity in the PCgG occur during eye closure and visual processing [38], and no significant change in blood flow to the PCgG has been shown between passive fixation and eye closure [39]. Therefore, it is very likely that the visual perceptual processing required in this experiment and the activity in this brain region are independent. Furthermore, the results of a study conducted by Uddin, Clare Kelly, Biswal, Castellanos, and Milham (2009), which showed a negative correlation between PCgG activity and the extrastriate visual areas involved in visual perceptual processing [40], suggest that it is very unlikely that the PCgG is involved in goal-directed visual perceptual information processing. Therefore, the result obtained in this study, i.e., the higher activation of the left PCgG in the picture observation condition than in the positive imagery generation condition, most likely reflects the perception-driven suppression of mental imagery generation rather than visual perceptual information processing.

If the left PCgG plays a role in suppressing the generation of any kind of mental imagery, then this implies that greater activity of the imagery suppression function, reflected in increased activity in the left PCgG, would lead to a reduction in the vividness of imagery. However, no such relationship was found in this experiment. This suggests that the perception- and emotion-driven suppression of imagery generation did not occur during the generation of positive imagery. Motoyama and Hishitani (2016) previously confirmed that the vividness of negative imagery varies with the activity of this brain area [24], and, consequently, the above fact suggests that the vividness of the positive imagery was influenced by other factors. It has been suggested that the vividness of imagery is not determined by a single function [1]. Therefore, future studies are needed to investigate how the vividness of positive imagery is determined, in addition to the imagery suppression function demonstrated in this study.

The left PCgG activity observed in this experiment may reflect other types of processing rather than the suppression of imagery generation. For example, it is likely that positive emotions were evoked in both the positive picture observation and positive imagery generation conditions. Although the participants' task in the positive picture observation condition was to observe positive pictures without interference, in the positive imagery generation condition, the participants simultaneously saw neutral pictures unrelated to the imagery object during positive imagery generation. That is, the evocation of positive emotions might be attenuated in the positive imagery generation condition compared to the case wherein positive imagery is simply generated without seeing a picture. A previous study also suggested that the PCgG is involved in the processing of positive emotions [41]. In other words, it could be considered that the activity in the left PCgG was stronger in the positive picture observation condition than in the positive imagery generation condition because the positive emotions evoked in the positive picture observation condition were greater than those evoked in the positive imagery generation condition despite the suppression of imagery generation. However, it is unlikely that the activity detected in the left PCgG in this experiment reflected positive emotion processing. It is likely that positive emotion processing occurred more in the picture observation and positive imagery generation conditions than in the baseline condition, which did not evoke emotions. That is, if PCgG activity reflected positive emotion processing, then this activity would also have been observed in the picture observation condition − in the baseline condition and in the positive imagery generation condition − in the baseline condition. In this experiment, no significant activity was found in either condition. These results suggest that the PCgG activity observed in this experiment does not reflect positive emotional processing.

5. Limitations and Implications of this Study

As the number of participants in this study may not be large enough, it would be desirable to conduct a similar experiment with a larger number of participants to see whether the results of this experiment can be replicated.

Furthermore, we believe that this study can be further improved by repeating similar experiments, not only with more participants but also with more accurate MRI equipment, such as 3T. The brain region responsible for perception-driven imagery suppression shown in the present study and the brain region responsible for emotion-driven imagery suppression shown in previous studies are both located in the left PCgG, but they do not completely overlap. In the future, it would be desirable to clarify whether the two functions are the same or different imagery suppression functions. By increasing the number of participants and using more accurate MRI machines to measure more precisely whether the brain regions activated during the two functions are different, it may be possible to investigate whether the perception-driven and emotion-driven functions are the same imagery suppression mechanism.

6. Conclusions

The novel finding of this study is the demonstration of the existence of a neural basis for the suppression of imagery generation, achieved using an approach that is different from that used in a previous study. In the previous study, significant activity in the left PCgG was found during negative imagery generation, where imagery generation was suppressed, compared to during positive imagery generation, where imagery generation was not suppressed. This finding suggests that the left PCgG is involved in the suppression of imagery generation. In this study, we investigated whether significant activity in the left PCgG also occurs during visual perceptual processing in which imagery generation is suppressed, which differed from the previous study. As a result, it was found that significant activity in the left PCgG occurred during visual perceptual processing compared to during positive imagery generation; thus, using a different approach to the previous study, this experiment provides further evidence that the left PCgG is involved in the suppression of imagery generation.

Author Contributions: Conceptualization, H.M. and S.H.; methodology, H.M. and S.H.; Software, H.M; formal analysis, H.M.; investigation, H.M.; data curation, H.M.; writing—original draft preparation, H.M.; writing—review and editing, S.H.; visualization, H.M.; supervision, S.H.; project administration, H.M. and S.H.; funding acquisition, H.M. and S.H. All authors have read and agreed to the published version of the manuscript.

Funding: This work was supported by JSPS KAKENHI Grant Number JP20K03477.

Institutional Review Board Statement: The study was approved by Hokkaido University, Graduate School of Letters, Ethics Committee (codes # 22nendo-1 and 23nendo-3).

Informed Consent Statement: Informed consent was obtained from all subjects involved in the study.

Data Availability Statement: The data used in this current study is available from the corresponding author upon reasonable request.

Acknowledgments: We would like to thank Fumihito Imai for his advice on data analysis.

Conflicts of Interest: The authors declare no conflicts of interest.

Appendix A

The pictures presented in the picture observation condition and the imagery object in the positive imagery generation condition. Half of the pictures were presented in the picture observation condition for one participant, and the words corresponding to these pictures were used as imagery objects in the positive imagery generation condition for the other participant. The eight pictures presented in the picture observation condition consisted of two 0 mm, two 4 mm, two 8 mm, two 12 mm of Gaussian filter effects.

The words used as the imagery objects and corresponding to the pictures	Gaussian filter effects			
	0mm	4mm	8mm	12mm
Bath				
Book				
Chocolate	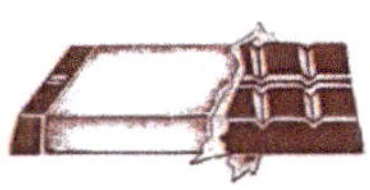			
Coast				

	Gaussian filtering effects			
	0mm	4mm	8mm	12mm
Fireplace				
Harp				
Kitten				
Peach	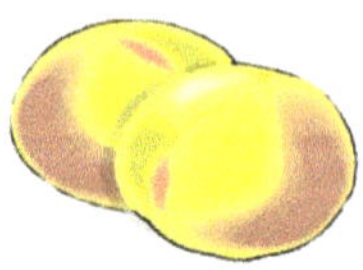		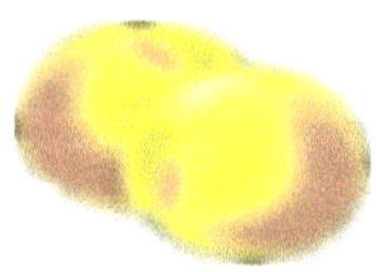	

	Gaussian filtering effects			
	0mm	4mm	8mm	12mm
Puppy				
Rabbit				
Rainbow				
Salad				

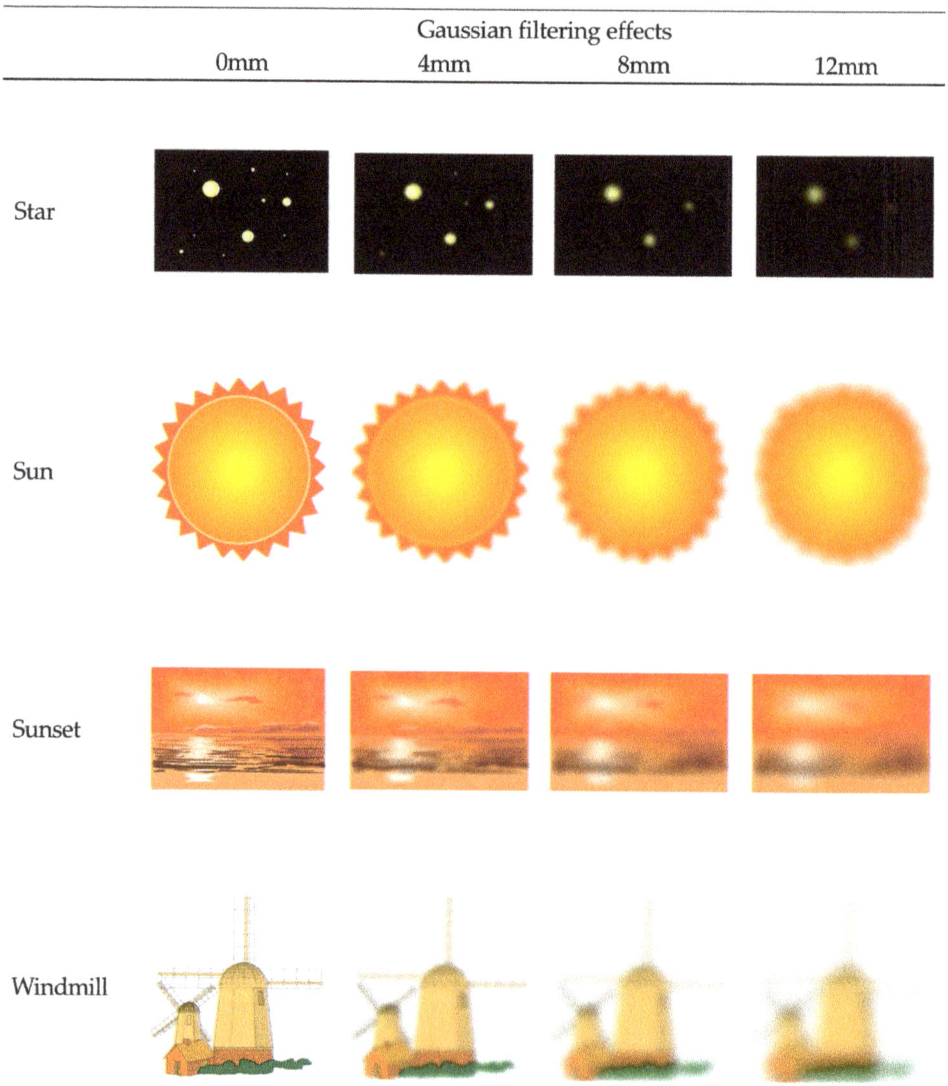

Appendix B

The pictures presented in the positive imagery generation condition. Half of the pictures were presented in one participant's positive imagery generation condition and the other half in the other participant's positive imagery generation condition. The eight pictures presented to each participant consisted of two 0 mm, two 4 mm, two 8 mm, two 12 mm of Gaussian filter effects.

Words corresponding to the pictures	Gaussian filter effects			
	0mm	4mm	8mm	12mm
Bus				
Car				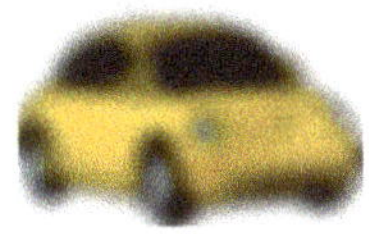
Chair				
Door				

	Gaussian filtering effects			
	0mm	4mm	8mm	12mm
Frypan				
Fork				
Fox				
Hat				

	Gaussian filtering effects			
	0mm	4mm	8mm	12mm
Kettle				
Microscope	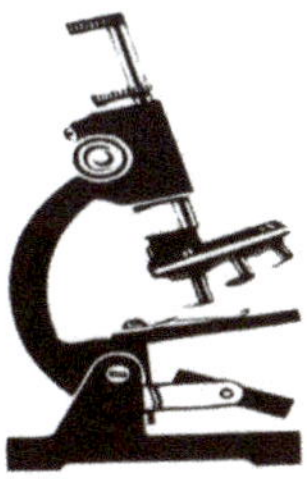			
Newspaper				
Pencil	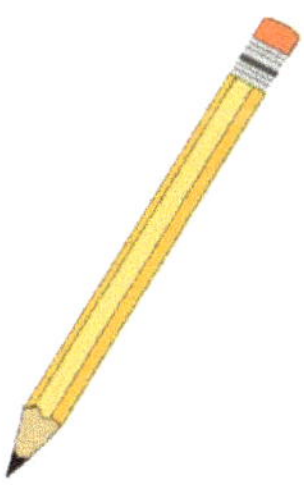			

	Gaussian filtering effects			
	0mm	4mm	8mm	12mm
Table				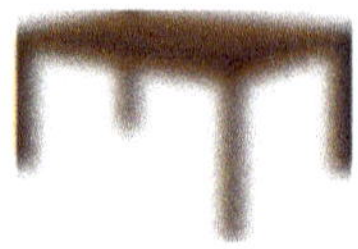
Telephone				
Thermometer	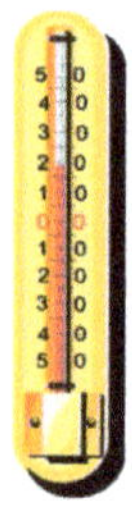	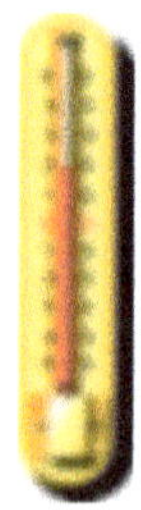		
Umbrella				

Notes

1 Previous research has not only shown that negative imageries are less vivid than positive ones, but also that negative autobiographical memories are less likely to be recalled. For example, involuntary autobiographical memories in everyday life have been shown to be more positive and neutral than those with negative emotions [42], and voluntary autobiographical memories are also more positive than negative, despite being weaker than involuntary autobiographical memories [43]. When participants in their 20s to 90s were asked about their involuntary autobiographical memories, it was found that they had more happy and mixed emotional memories than sad ones [44].

2 Intrusive imagery, a type of negative imagery that is a clinical symptom of PTSD flashbacks, depression, and anxiety, has been shown to be extremely vivid [45]. Motoyama and Hishitani (2016) described negative imagery as being more vague than positive imagery [24]. However, the negative imagery that they focused on was not intrusive imagery but rather imagery that was intentionally generated by healthy individuals. Therefore, the vague negative imagery described by Motoyama and Hishitani (2016) does not include intrusive imagery.

[3] In this experiment, we used the term "vividness" when participants rated the vividness of the generated imageries in the positive imagery condition, and we used the term "richness" when participants rated the clarity of the presented picture in the picture observation condition.

References

1. Hishitani, S.; Miyazaki, T.; Motoyama, H. Some mechanisms responsible for the vividness of mental imagery: Suppressor, closer, and other functions. *J. Ment. Imag.* **2011**, *35*, 5–32. Available online: http://hdl.handle.net/2115/47716 (accessed on 8 January 2024).
2. Kosslyn, S.M. *Image and Mind*; Harvard University Press: Cambridge, MA, USA, 1980.
3. Kosslyn, S.M.; Thompson, W.L.; Ganis, G. *The Case for Mental Imagery*; Oxford University Press: New York, NY, USA, 2006. [CrossRef]
4. Pearson, D.G. Imagery and the visuo-spatial sketchpad. In *Working Memory in Perspective*; Andrade, J., Ed.; Psychology Press: London, UK, 2001; pp. 33–59.
5. Pearson, D.G.; Logie, R.H.; Gilhooly, K.J. Verbal representation and spatial manipulation during mental synthesis. *Eur. J. Cogn. Psychol.* **1999**, *11*, 295–314. [CrossRef]
6. Pearson, D.G.; Logie, R.H.; Green, C. Mental manipulation, visual working memory, and executive processes. *Psychol. Beitr.* **1996**, *38*, 324–342.
7. Kosslyn, S.M. *Image and Brain: The Resolution of the Imagery Debate*; MIT Press: Cambridge, MA, USA, 1994. [CrossRef]
8. Brooks, L.R. The suppression of visualization by reading. *Q. J. Exp. Psychol.* **1967**, *19*, 289–299. [CrossRef]
9. Chang, S.; Lewis, D.E.; Pearson, J. The functional effects of color perception and color imagery. *J. Vis.* **2013**, *13*, 1168. [CrossRef]
10. Craver-Lemley, C.; Arterberry, M.; Reeves, A. Effects of imagery on vernier acuity under conditions of induced depth. *J. Exp. Psychol. Hum. Percept. Perform.* **1997**, *23*, 3–13. [CrossRef]
11. Craver-Lemley, C.; Arterberry, M.; Reeves, A. "Illusory" illusory conjunctions: The conjoining of features of visual and imagined stimuli. *J. Exp. Psychol. Hum. Percept. Perform.* **1999**, *25*, 1036–1049. [CrossRef]
12. Craver-Lemley, C.; Reeves, A. Visual imagery selectively reduces vernier acuity. *Perception* **1987**, *16*, 599–614. [CrossRef]
13. Craver-Lemley, C.; Reeves, A. How visual imagery interferes with vision. *Psychol. Rev.* **1992**, *99*, 633–649. [CrossRef]
14. Ishai, A.; Sagi, D. Visual imagery: Effects of short- and long-term memory. *J. Cogn. Neurosci.* **1997**, *9*, 734–742. [CrossRef]
15. Keogh, R.; Pearson, J. Mental imagery and visual working memory. *PLoS ONE* **2011**, *6*, e29221. [CrossRef] [PubMed]
16. Keogh, R.; Pearson, J. The sensory strength of voluntary visual imagery predicts visual working memory capacity. *J. Vis.* **2014**, *14*, 7. [CrossRef] [PubMed]
17. Keogh, R.; Pearson, J. The perceptual and phenomenal capacity of mental imagery. *Cognition* **2017**, *162*, 124–132. [CrossRef] [PubMed]
18. Kwok, E.L.; Leys, G.; Koenig-Robert, R.; Pearson, J. Measuring Thought-Control Failure: Sensory Mechanisms and Individual Differences. *Psychol. Sci.* **2019**, *30*, 811–821. [CrossRef] [PubMed]
19. Reeves, A. Visual imagery lowers sensitivity to hue-varying, but not to luminance-varying, visual stimuli. *Percept. Psychophys.* **1981**, *29*, 247–250. [CrossRef] [PubMed]
20. Reeves, A.; Craver-Lemley, C. Unmasking the Perky effect: Spatial extent of image interference on visual acuity. *Front. Psychol.* **2012**, *3*, 296. [CrossRef] [PubMed]
21. Segal, S.J.; Fusella, V. Influence of imaged pictures and sounds on detection of visual and auditory signals. *J. Exp. Psychol.* **1970**, *83*, 458–464. [CrossRef] [PubMed]
22. Sherwood, R.; Pearson, J. Closing the mind's eye: Incoming luminance signals disrupt visual imagery. *PLoS ONE* **2010**, *5*, e15217. [CrossRef] [PubMed]
23. Kosslyn, S.M. Seeing and imaging in the cerebral hemispheres: A computational approach. *Psychol. Rev.* **1987**, *94*, 148–175. [CrossRef]
24. Motoyama, H.; Hishitani, S. The brain mechanism that reduces the vividness of negative imagery. *Conscious. Cogn.* **2016**, *39*, 59–69. [CrossRef]
25. Bywaters, M.; Andrade, J.; Turpin, G. Intrusive and non-intrusive memories in a non-clinical sample: The effects of mood and affect on imagery vividness. *Memory* **2004**, *12*, 467–478. [CrossRef] [PubMed]
26. Hertel, P.T.; Parks, C. Emotional episodes facilitate word recall. *Cogn. Emot.* **2002**, *16*, 685–694. [CrossRef]
27. Hishitani, S. Toward a deeper understanding of vividness: Some points inspired from McKelvie's Article. *J. Ment. Imag.* **1995**, *19*, 139–143.
28. Segal, S.J.; Fusella, V. Effect of images in six sense modalities on detection of visual signal from noise. *Psychon. Sci.* **1971**, *24*, 55–56. [CrossRef]
29. Oldfield, R.C. The assessment and analysis of handedness: The Edinburgh inventory. *Neuropsychologia* **1971**, *9*, 97–113. [CrossRef] [PubMed]
30. Miyazaki, T.; Motoyama, H.; Hishitani, S. Emotional values for nouns and epithets: Standardization on the positive-negative dimension. *Jpn. J. Ment. Imag.* **2003**, *1*, 48–59. Available online: http://hdl.handle.net/2115/44490 (accessed on 8 April 2024).
31. Nova Development. *Art Explosion 750,000 for Macintosh*; Nova Development: Calabasas, CA, USA, 2003; Available online: http://usa.novadevelopment.com/company/reseller_center/ArtExplosion750000.aspx (accessed on 8 April 2024).

32. Hishitani, S.; Motoyama, H. The trial manufacture of switch box for fMRI experiment. In *Hardware and Software for Psychology Laboratories*; Hishitani, S., Ed.; Nakanishiya Shuppan: Kyoto, Japan, 2009; pp. 203–211.
33. Maldjian, J.A.; Laurienti, P.J.; Burdette, J.H. Precentral gyrus discrepancy in electronic versions of the Talairach atlas. *NeuroImage* **2004**, *21*, 450–455. [CrossRef]
34. Maldjian, J.A.; Laurienti, P.J.; Kraft, R.A.; Burdette, J.B. An automated method for neuroanatomic and cytoarchitectonic atlas-based interrogation of fMRI data sets. *NeuroImage* **2003**, *19*, 1233–1239. [CrossRef]
35. Tzourio-Mazoyer, N.; Landeau, B.; Papathanassiou, D.; Crivello, F.; Etard, O.; Delcroix, N.; Mazoyer, B.; Joliot, M. Automated anatomical labeling of activations in SPM using a macroscopic anatomical parcellation of the MNI MRI single-subject brain. *Neuroimage* **2002**, *15*, 273–289. [CrossRef]
36. Brett, M.; Anton, J.L.; Valabregue, R.; Poline, J.B. Region of interest analysis using an SPM toolbox. In Proceedings of the 8th International Conference on Functional Mapping of the Human Brain, Sendai, Japan, 2–6 June 2002.
37. Buckner, R.L.; Andrews-Hanna, J.R.; Schacter, D.L. The brain's default network: Anatomy, function, and relevance to disease. *Ann. New York Acad. Sci.* **2008**, *1124*, 1–38. [CrossRef]
38. Greicius, M.D.; Krasnow, B.; Reiss, A.L.; Menon, V. Functional connectivity in the resting brain: A network analysis of the default mode hypothesis. *Proc. Natl. Acad. Sci. USA* **2003**, *100*, 253–258. [CrossRef] [PubMed]
39. Raichle, M.E.; MacLeod, A.M.; Snyder, A.Z.; Powers, W.J.; Gusnard, D.A.; Shulman, G.L. A default mode of brain function. *Proc. Natl. Acad. Sci. USA* **2001**, *98*, 676–682. [CrossRef] [PubMed]
40. Uddin, L.Q.; Clare Kelly, A.M.; Biswal, B.B.; Castellanos, F.X.; Milham, M.P. Functional connectivity of default mode network components: Correlation, anticorrelation, and causality. *Hum. Brain Mapp.* **2009**, *30*, 625–637. [CrossRef] [PubMed]
41. Aldhafeeri, F.M.; Mackenzie, I.; Kay, T.; Alghamdi, J.; Sluming, V. Regional brain responses to pleasant and unpleasant IAPS pictures: Different networks. *Neurosci. Lett.* **2012**, *512*, 94–98. [CrossRef] [PubMed]
42. Berntsen, D. Involuntary autobiographical memories. *Appl. Cogn. Psychol.* **1996**, *10*, 434–454. [CrossRef]
43. Berntsen, D. Voluntary and involuntary access to autobiographical memory. *Memory* **1998**, *6*, 113–141. [CrossRef] [PubMed]
44. Berntsen, D.; Rubin, D.C. Emotionally Charged Autobiographical Memories Across the Life Span: The Recall of Happy, Sad, Traumatic, and Involuntary Memories. *Psychol. Aging* **2002**, *17*, 636–652. [CrossRef]
45. Çili, S.; Stopa, L. Intrusive mental imagery in psychological disorders: Is the self the key to understanding maintenance? *Front. Psychiatry* **2015**, *6*, 103. [CrossRef]

vision

MDPI

Review

Uncovering the Role of the Early Visual Cortex in Visual Mental Imagery

Nadine Dijkstra

Department of Imaging Neuroscience, Institute of Neurology, University College London, London WC1E 6BT, UK; n.dijkstra@ucl.ac.uk

Abstract: The question of whether the early visual cortex (EVC) is involved in visual mental imagery remains a topic of debate. In this paper, I propose that the inconsistency in findings can be explained by the unique challenges associated with investigating EVC activity during imagery. During perception, the EVC processes low-level features, which means that activity is highly sensitive to variation in visual details. If the EVC has the same role during visual mental imagery, any change in the visual details of the mental image would lead to corresponding changes in EVC activity. Within this context, the question should not be whether the EVC is 'active' during imagery but how its activity relates to specific imagery properties. Studies using methods that are sensitive to variation in low-level features reveal that imagery can recruit the EVC in similar ways as perception. However, not all mental images contain a high level of visual details. Therefore, I end by considering a more nuanced view, which states that imagery can recruit the EVC, but that does not mean that it always does so.

Keywords: mental imagery; visual perception; early visual cortex

1. Introduction

Close your eyes and imagine an apple. Is this experience similar to actually seeing an apple or is it more like thinking about the idea of an apple? Answering this question, whether imagery relies on depictive or symbolic representations, has been referred to as the 'imagery-debate'. This debate has dominated mental imagery research during the end of the last century and the beginning of this one [1,2]. The development of cognitive neuroimaging promised a way to settle the debate once and for all: if mental images, like perceived images, are represented in the retinotopically organized early visual cortex (EVC), this would be knock-out evidence in favor of a depictive view. However, at this moment, more than three decades after the first neuroimaging studies on mental imagery, the role of the early visual cortex during visual mental imagery remains a topic of intense debate [3–6].

In this paper, I argue that in order to make progress, we need to move away from the question of *whether* the EVC is involved in mental imagery and instead move towards elucidating exactly *how* it is involved. I argue that many of the discrepancies in the literature can be explained by the unique methodological and conceptual challenges associated with the functional role of the EVC. I will start by explaining that the question of whether the EVC is 'active' or not during imagery is conceptually nonsensical in light of what we know about the functionality of the EVC. Instead, to elucidate the involvement of the EVC during mental imagery, its unique organization needs to be considered both in the experimental design as well as in the data analysis. I will discuss examples of studies that do this and explain what they tell us about how the EVC is involved in visual mental imagery. I focus on the volitional use of mental imagery and will mostly discuss studies in which participants are explicitly instructed to form mental images. In Section 4, I will further discuss to what extent these conclusions generalize to more implicit forms of imagery.

Taken together, I propose that the current evidence suggests that the EVC represents fine-grained visual details during mental imagery in the same way as it does during perception. I will end by discussing how this account can explain contradictory evidence and highlight directions for future research.

2. Organizational Principles of the Early Visual Cortex

Decades of neurophysiological research have led to a highly detailed characterization of how the early visual cortex responds to external visual signals. Activity in the EVC is retinotopically organized such that things that are spatially close together in the outside world elicit activity in neurons that are close together on the cortical sheet of the EVC. Furthermore, neurons in the EVC have small receptive fields so that each neuron only responds to signals coming from a very small portion of the visual field. Together, this means that activity in the EVC is highly sensitive to changes in low-level features of visual signals, such as their exact distribution of edges, location and orientation [7]. This is in contrast to neurons in more high-level visual areas that have much larger receptive fields and are therefore able to represent more 'high-level' features, such as the semantic meaning of a stimulus, irrespective of the specific low-level instantiation of that stimulus [8]. It is important to note that this distinction between low- and high-level sensitivity is not discrete. Specifically, activity in the EVC has also been shown to be modulated by high-level properties such as semantic scene context [9,10], and activity in high-level cortex has also been shown to be influenced by certain low-level features such as location [11]. However, in general, activity in the EVC is much more sensitive to variation in low-level features [12]. This means that, for example, the same neurons in the high-level visual cortex might fire in response to a chair from different viewpoints, whereas the pattern of activity in the EVC will be widely different in each specific instance.

The properties of the neurons in the EVC have the consequence that any change in low-level features of visual signals, for example, the exact orientation of a stimulus, will lead to a corresponding change in EVC activity (Figure 1A). While looking outside your window at a tree, any displacement of the branches or the leaves due to the wind will result in a change in activity in your early visual cortex. Furthermore, any movement of your eyes will change in which receptive fields the visual signals of the tree fall, again leading to corresponding changes in EVC activity. If you were to now test whether any given neuron in the EVC was consistently active when you were looking at this tree, you would likely find very few neurons that showed consistent activity. Based on this, you could wrongfully conclude that the EVC was not involved in the perception of the tree. This is why most visual neuroscience experiments studying the EVC are done with controlled repetitions of the exact same set of simple stimuli, presented in exactly the same location on the screen, with the participant's head firmly fixated on a chin rest and taking dedicated measures to minimize eye-movements as much as possible. Under these controlled conditions, it is much more likely to observe consistent activity in EVC neurons.

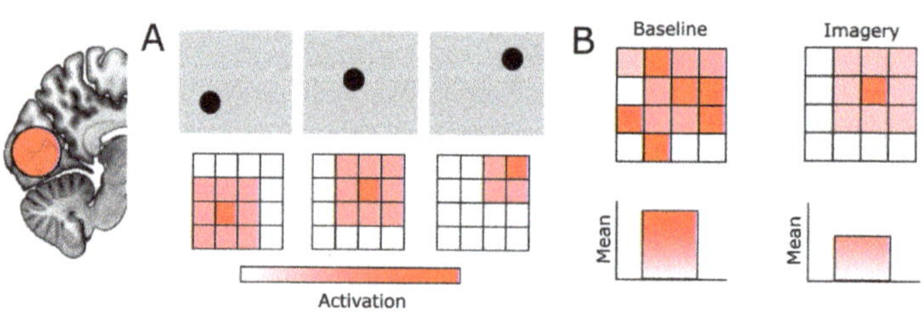

Figure 1. Challenging properties of early visual cortex (EVC) activity during imagery. (**A**) Activity in the EVC is highly sensitive to changes in low-level features, such as where in visual space a stimulus is located. Top row indicates toy visual signals, spreading over the visual field. Bottom row indicates the activity pattern in a grid of EVC neurons. (**B**) Imagery is thought to be instantiated through inhibitory feedback connections, leading to the sharpening of representations rather than increases in general activation levels. This means that the EVC can still be involved in imagery even if there is no change or even a decrease in general activation.

3. The Challenges of Studying the Early Visual Cortex during Mental Imagery

Now, consider imagining a tree in your mind's eye. How stable are the low-level features of this mental image? Are the locations of the branches fixed and clear from one moment to the next? Are you able to hold a clear and stable image without moving your eyes? A recent study showed that when asked to freely imagine a scene, most participants leave out many low-level details [13]. Furthermore, some people might even experience mental images in a kind of separate space that is decoupled from the physical environment [14]. Herein lies the main difficulty of studying the role of the EVC in imagery: Given the sensitivity of activity in the EVC to changes in low-level features, the fleeting and undefined nature of mental images makes it very hard to determine any consistent relationship. In order to properly study the role of the EVC in imagery, it is therefore essential to have a firm handle on the low-level features of the mental images under investigation.

One way to control the low-level features of mental images is by instructing participants to imagine very simple stimuli. One of the earliest studies using this approach was by Klein and colleagues in 2004 [15]. In this study, participants were instructed to view and imagine horizontally and vertically oriented bow-tie stimuli while their brain activity was measured using functional magnetic resonance imaging (fMRI). Crucially, the exact spatial location and orientation of the to-be-imagined stimuli were clearly defined. The results revealed that imagining a horizontal stimulus activated the horizontal meridian within the EVC while imagining a vertical stimulus activated the vertical meridian, in line with retinotopic involvement of the EVC during imagery [15]. Similar approaches have been used to show retinotopic encoding in the EVC of imagined domino patterns [16] and letters [17].

Another unique challenge of studying the EVC during imagery is that the way that EVC activity is instantiated during imagery is different from perception. Specifically, during imagery, activity in the EVC is assumed to be instantiated via feedback connections from high-level areas [18–20]. Importantly, in contrast to feedforward connections, feedback connections tend to predominantly *modulate* neural activity, changing existing firing rates via gain control, but usually without driving neurons to fire action potentials directly [21,22]. In line with this, it has recently been proposed that imagery inhibits activity in irrelevant neuronal populations rather than directly increasing activity in the populations representing the imagined stimulus [23]. This is similar to the 'sharpening' process assumed to underlie attention [24] and expectation [25]. Instead of modulating bottom-up signals like during attention and expectation, imagery is assumed to modulate spontaneous, baseline fluctuations in brain activity, carving out stimulus representations by inhibiting irrelevant activity [5,26]. Importantly, if imagery is instantiated through inhibition of irrelevant activity, overall activity levels in the EVC might not increase and might even decrease compared to baseline (Figure 1B). This could explain the observation that aphantasia—the absence of visual mental imagery—has been associated with hyperactivity of the EVC [27,28]. Importantly, however, if the EVC is involved in imagery, the relative activity pattern should still be informative of which stimulus participants were imagining, even if general activation levels do not increase.

Testing whether neural activity patterns contain information about perceived or imagined stimuli is the main goal of 'decoding' techniques in cognitive neuroscience. In this context, decoding refers to the use of machine learning algorithms to describe how neural activity patterns relate to different stimuli or conditions [29]. In a standard imagery decoding experiment, a decoder is first 'trained' to capture the variation in brain activity patterns during the perception of different stimuli. This perception-decoder is then applied to the brain activity during imagery of the same stimuli, giving a 'guess' of which stimulus the participant was imagining. If the perception-classifier is able to accurately guess or 'decode' the imagined stimulus, it can be concluded that the pattern of neural activity during imagery is similar to that during perception in that brain area. Using this technique, significant cross-decoding between imagery and perception in the EVC has been shown for gratings, letters, objects and shapes [17,30–32].

Furthermore, to more directly test whether low-level visual features are encoded in EVC activity patterns, researchers have used the so-called feature encoding models [33]. These are models based on computer vision algorithms that describe complex images in terms of unique combinations of simple, low-level features. These models can, in turn, be combined with models that describe activity in different brain areas in response to the same low-level features to predict what the brain activity would be like for an entirely new set of complex images [34]. For example, one study used a low-level feature encoding model to first determine how images of complex artworks could be described in terms of a combination of Gabor wavelets [35]. Next, the activity in the EVC during the perception of the same Gabor wavelets was measured to create a corresponding low-level feature model of EVC activity. Applying the transformation between Gabor wavelets and the artworks to EVC activity during the imagery of different artworks led to the successful prediction of which artwork participants were imagining [35]. Several other recent studies have used similar computer vision approaches to reveal low-level feature encoding of visual mental imagery in brain activity patterns [36,37].

Together, these studies highlight that in order to capture EVC involvement during visual mental imagery, it is essential to use the right experimental and analytical approach. In this context, investigating whether the EVC is 'active' or not without specifying the exact way in which its activity relates to specific properties of the mental image is unlikely to reveal consistent involvement. For example, consider an experiment in which, on each trial, participants are asked to imagine a specific stimulus but are not asked to imagine that stimulus in a specific size or at a specific location. Due to the organization of the EVC, each of those imagery instances would be associated with different patterns of activity. Averaging over trials might, therefore, lead to the incorrect conclusion that imagery did not recruit the EVC. This can also explain why meta-analyses investigating the activation of imagery of different stimuli over different studies are ill-placed to find activation in the EVC and are instead more likely to find consistent involvement of higher-level visual regions [6].

4. The Role of the Early Visual Cortex in Visual Mental Imagery

Considering the evidence until now, I propose that the EVC represents fine-grained visual details during imagery in similar ways as it does during perception, in line with a depictive view. However, importantly, that does not mean that all instances of visual mental imagery will activate the EVC. Consider, for example, a recent study in which participants were asked to imagine a person walking into a room and knocking a ball off a table [13]. Afterwards, participants were asked several questions regarding the visual details of their mental imagery, i.e., 'did you imagine the color of the ball', 'the clothes the person was wearing', etc. It turned out that most participants did not imagine the majority of basic visual details. The authors concluded that 'while imagination may indeed be a good artist, it's on a deadline, and stingy about paint' ([13], p. 20). In other words, imagination *can* create vivid, detailed scenes, but that requires effort, so if it is not necessary, those details will be omitted.

This means that EVC activity is likely not always observed during mental imagery, but mostly when low-level details are indeed imagined. The idea that EVC activity depends on whether the imagery in question requires fine-grained details is in line with early observations that noted that the EVC is more likely to be activated in imagery tasks that require high-resolution discrimination judgments [38]. Furthermore, the observation that the vividness of mental imagery correlates with perception-imagery cross-decoding accuracy in EVC [31,39,40] is also in line with the idea that more detailed mental images are more likely to recruit the EVC.

One line of research that has been taken as evidence against the involvement of the EVC in visual mental imagery comes from neuropsychological observations of intact visual mental imagery after lesions to the EVC [3,41]. One way to align these findings with the previously mentioned neuroimaging research is with the fact that imagery does not always contain fine-grained visual details [42]. This would mean that patients without a functioning EVC might still be able to generate mental images, but that these mental images would not contain the fine-grained visual details that require a functioning EVC. Future studies using controlled psychophysical paradigms in these patients are required to determine whether this is indeed the case.

Another potential argument against the involvement of the EVC in imagery comes from recent studies that show that imagery can be decoded from EVC activity in aphantasia [43–45]. However, the claim that EVC representations are necessary to imagine low-level features does not mean that these representations are also sufficient for conscious visual imagery. One possibility is that these low-level EVC representations need to be coupled with higher-level representations in frontoparietal areas in order to reach awareness, as proposed by higher-order theories of consciousness [46–48]. In line with this, a recent study found that there were no differences in visual regions between aphantasia and controls but that in aphantasia, visual regions were functionally disconnected from the fronto-parietal cortex [49].

The EVC also seems to represent internally generated visual information in contexts where imagery is not explicitly instructed. For example, EVC activity tracks the generated stimulus during mental rotation [30,50] as well as during visual working memory [30,51], even though both of these processes have been shown to be largely unaffected in aphantasia [52,53], suggesting that they might not require conscious mental imagery. These findings suggest that the EVC represents internally generated visual information irrespective of whether these signals are instantiated through volitional, conscious imagery or not.

5. Conclusions and Future Directions

The question of whether visual mental imagery recruits the early visual cortex has been at the core of the imagery-debate for decades. Initial neuroimaging findings were inconclusive, with some studies finding reliable activity while other studies did not. In this paper, I have suggested that the inconsistency in the literature can be explained by the unique challenges associated with studying the EVC during visual mental imagery. The EVC is known for representing fine-grained visual details during perception. However, compared to perception, imagery tends to be associated with weaker and less consistent visual signals. Moreover, some instances of mental imagery might contain very few fine-grained visual details at all. These challenges require specific experimental and analytical approaches, such as using very simple stimuli or quantifying expected activity patterns with sophisticated models. The results of studies using these approaches consistently show that the EVC represents low-level visual details during mental imagery in similar ways as it does during perception.

A number of questions regarding the exact role of the EVC in visual mental imagery remain open. Firstly, the exact relationship between EVC representations and conscious imagery experience remains unclear. Several studies have shown a relationship between EVC involvement and subjective reports of imagery vividness [31,39,40]. However, recent studies suggest that EVC representations can be present in the complete absence of conscious visual imagery [43–45]. Future research is necessary to investigate which factors determine whether imagery-related EVC activity becomes conscious [54].

Relatedly, exactly how visual mental imagery representations are instantiated in the EVC remains unclear. The current consensus is that this happens through (inhibitory) feedback connections from high-level areas [4,5]. However, it is not clear exactly which high-level regions are referred to and exactly how activation flows through these different regions. Several studies have found increases in coupling between the dorsolateral prefrontal and visual cortex during imagery [19,20]. Other studies have shown increases in coupling between parietal and visual regions [18]. Lastly, research on scene imagery has focused on the role of medial frontal and hippocampal memory systems in generating mental images [55]. Further research is necessary to fully elucidate the neural network involved in generating EVC activity during mental imagery.

Finally, given that imagery seems to be able to recruit the EVC in a similar way to perception, one important open question is how the brain keeps track of which visual signals reflect external reality and which reflect internal imagination. One hypothesis has been that the distinction is made based on whether the middle layer of the EVC is involved, which is where bottom-up sensory signals arrive in the cortex [56]. Another hypothesis is that the feeling of 'realness' is determined by whether or not the superficial layers of the EVC are activated [57]. Initial results show that during mental imagery, only deep layers of the EVC are involved. In contrast, superficial layers do become activated during seemingly 'real' illusory perception [57]. Another not mutually exclusive hypothesis is that whether EVC signals are considered real or not depends on whether they are strong enough to cross a 'reality threshold' [40]. Finally, other proposals have suggested that external, feedforward and internal feedback signals might be separated in time, where feedforward signals might be carried by faster oscillations in the gamma band, while feedback signals are carried by slower oscillations in the alpha/beta band [58–62]. Further elucidating how the brain keeps apart EVC signals of imagery and perception has important implications for our understanding of when this line breaks down in hallucinations.

To conclude, developments in both experimental design and analytical approaches are converging on the idea that the EVC is recruited when low-level visual details are imagined. Future research is needed to further characterize exactly how activity in the EVC is instantiated during imagery and how this activity leads to the unique conscious experience characteristic of our imagination. Answering these remaining questions will further increase our fundamental understanding of the neural correlates of consciousness as well as provide novel perspectives on understanding disorders of mental imagery.

Funding: This research was funded by a Marie Curie grant from the European Union Horizon 2020 program [882832/MSCA/IF/EF/ST], UK Research and Innovation (UKRI) under the UK government's Horizon Europe funding guarantee [selected as ERC Consolidator, grant number 101043666] and an Adversarial Collaboration Grant from the Templeton World Charity Foundation [TWCF number 22032].

Acknowledgments: The author would also like to thank Benjy Barnett for his comments on an earlier version of this manuscript.

Conflicts of Interest: The author declares no conflict of interest.

References

1. Nanay, B. Mental Imagery. In *The Stanford Encyclopedia of Philosophy*; Winter 2021; Zalta, E.N., Ed.; Metaphysics Research Lab, Stanford University: Stanford, CA, USA, 2021. Available online: https://plato.stanford.edu/archives/win2021/entries/mental-imagery/ (accessed on 27 July 2023).
2. Sterelny, K. The Imagery Debate. *Philos. Sci.* **1986**, *53*, 560–583. [CrossRef]
3. Bartolomeo, P.; Hajhajate, D.; Liu, J.; Spagna, A. Assessing the causal role of early visual areas in visual mental imagery. *Nat. Rev. Neurosci.* **2020**, *21*, 517. [CrossRef] [PubMed]
4. Dijkstra, N.; Bosch, S.E.; van Gerven, M.A.J. Shared Neural Mechanisms of Visual Perception and Imagery. *Trends Cogn. Sci.* **2019**, *23*, 423–434. [CrossRef] [PubMed]
5. Pearson, J. The human imagination: The cognitive neuroscience of visual mental imagery. *Nat. Rev. Neurosci.* **2019**, *20*, 624–634. [CrossRef] [PubMed]
6. Spagna, A.; Hajhajate, D.; Liu, J.; Bartolomeo, P. Visual mental imagery engages the left fusiform gyrus, but not the early visual cortex: A meta-analysis of neuroimaging evidence. *Neurosci. Biobehav. Rev.* **2021**, *122*, 201–217. [CrossRef] [PubMed]
7. Hubel, D.H.; Wiesel, T.N. Receptive Fields and Functional Architecture of monkey striate cortex. *J. Physiol.* **1968**, *195*, 215–243. [CrossRef] [PubMed]
8. Thorpe, S.J.; Fabre-Thorpe, M. Seeking categories in the brain. *Science* **2001**, *291*, 260–263. [CrossRef] [PubMed]
9. Morgan, A.T.; Petro, L.S.; Muckli, L. Scene representations conveyed by cortical feedback to early visual cortex can be described by line drawings. *J. Neurosci.* **2019**, *39*, 9410–9423. [CrossRef]
10. Muckli, L.; De Martino, F.; Vizioli, L.; Petro, L.S.; Smith, F.W.; Ugurbil, K.; Goebel, R.; Yacoub, E. Contextual Feedback to Superficial Layers of V1. *Curr. Biol.* **2015**, *25*, 2690–2695. [CrossRef]
11. Kravitz, D.J.; Kriegeskorte, N.; Baker, C.I. High-Level Visual Object Representations Are Constrained by Position. *Cereb. Cortex* **2010**, *20*, 2916–2925. [CrossRef]

12. Park, S.; Serences, J.T. Relative precision of top-down attentional modulations is lower in early visual cortex compared to mid- and high-level visual areas. *J. Neurophysiol.* **2022**, *127*, 504–518. [CrossRef]
13. Bigelow, E.J.; McCoy, J.P.; Ullman, T.D. Non-commitment in mental imagery. *Cognition* **2023**, *238*, 105498. [CrossRef]
14. Schwarzkopf, D.S. What is the true range of mental imagery? *Cortex* **2024**, *170*, 21–25. [CrossRef] [PubMed]
15. Klein, I.; Dubois, J.; Mangin, J.-F.; Kherif, F.; Flandin, G.; Poline, J.-B.; Denis, M.; Kosslyn, S.M.; Le Bihan, D. Retinotopic organization of visual mental images as revealed by functional magnetic resonance imaging. *Brain Res. Cogn. Brain Res.* **2004**, *22*, 26–31. [CrossRef]
16. Thirion, B.; Duchesnay, E.; Hubbard, E.; Dubois, J.; Poline, J.-B.; Lebihan, D.; Dehaene, S. Inverse retinotopy: Inferring the visual content of images from brain activation patterns. *NeuroImage* **2006**, *33*, 1104–1116. [CrossRef] [PubMed]
17. Senden, M.; Emmerling, T.C.; van Hoof, R.; Frost, M.A.; Goebel, R. Reconstructing imagined letters from early visual cortex reveals tight topographic correspondence between visual mental imagery and perception. *Brain Struct. Funct.* **2019**, *224*, 1167–1183. [CrossRef] [PubMed]
18. Dentico, D.; Cheung, B.L.; Chang, J.-Y.; Guokas, J.; Boly, M.; Tononi, G.; Van Veen, B. Reversal of cortical information flow during visual imagery as compared to visual perception. *NeuroImage* **2014**, *100*, 237–243. [CrossRef]
19. Dijkstra, N.; Zeidman, P.; Ondobaka, S.; van Gerven, M.A.J.; Friston, K. Distinct Top-down and Bottom-up Brain Connectivity During Visual Perception and Imagery. *Sci. Rep.* **2017**, *7*, 5677. [CrossRef] [PubMed]
20. Mechelli, A.; Price, C.J.; Friston, K.J.; Ishai, A. Where bottom-up meets top-down: Neuronal interactions during perception and imagery. *Cereb. Cortex* **2004**, *14*, 1256–1265. [CrossRef]
21. Aru, J.; Siclari, F.; Phillips, W.A.; Storm, J.F. Apical drive—A cellular mechanism of dreaming? *Neurosci. Biobehav. Rev.* **2020**, *119*, 440–455. [CrossRef]
22. Larkum, M. A cellular mechanism for cortical associations: An organizing principle for the cerebral cortex. *Trends Neurosci.* **2013**, *36*, 141–151. [CrossRef]
23. Pace, T.; Koenig-Robert, R.; Pearson, J. Different Mechanisms for Supporting Mental Imagery and Perceptual Representations: Modulation Versus Excitation. *Psychol. Sci.* **2023**, *34*, 1229–1243. [CrossRef]
24. Bartsch, M.V.; Loewe, K.; Merkel, C.; Heinze, H.-J.; Schoenfeld, M.A.; Tsotsos, J.K.; Hopf, J.-M. Attention to Color Sharpens Neural Population Tuning via Feedback Processing in the Human Visual Cortex Hierarchy. *J. Neurosci.* **2017**, *37*, 10346–10357. [CrossRef]
25. Kok, P.; Jehee, J.F.M.; de Lange, F.P. Less Is More: Expectation Sharpens Representations in the Primary Visual Cortex. *Neuron* **2012**, *75*, 265–270. [CrossRef]
26. Koenig-Robert, R.; Pearson, J. Why do imagery and perception look and feel so different? *Philos. Trans. R. Soc. B Biol. Sci.* **2021**, *376*, 20190703. [CrossRef]
27. Keogh, R.; Bergmann, J.; Pearson, J. Cortical excitability controls the strength of mental imagery. *eLife* **2020**, *9*, e50232. [CrossRef]
28. Winlove, C.; Milton, F.; Ranson, J.; Fulford, J.; MacKisack, M.; Macpherson, F.; Zeman, A. The neural correlates of visual imagery: A co-ordinate-based meta-analysis. *Cortex* **2018**, *105*, 4–25. [CrossRef]
29. Robinson, A.K.; Quek, G.L.; Carlson, T.A. Visual Representations: Insights from Neural Decoding. *Annu. Rev. Vis. Sci.* **2023**, *9*, 313–335. [CrossRef]
30. Albers, A.M.; Kok, P.; Toni, I.; Dijkerman, H.C.; De Lange, F.P. Shared representations for working memory and mental imagery in early visual cortex. *Curr. Biol. CB* **2013**, *23*, 1427–1431. [CrossRef]
31. Lee, S.-H.; Kravitz, D.J.; Baker, C.I. Disentangling visual imagery and perception of real-world objects. *NeuroImage* **2012**, *59*, 4064–4073. [CrossRef]
32. Ragni, F.; Tucciarelli, R.; Andersson, P.; Lingnau, A. Decoding stimulus identity in occipital, parietal and inferotemporal cortices during visual mental imagery. *Cortex* **2020**, *127*, 371–387. [CrossRef]
33. St-Yves, G.; Naselaris, T. The feature-weighted receptive field: An interpretable encoding model for complex feature spaces. *NeuroImage* **2018**, *180*, 188–202. [CrossRef]
34. Guclu, U.; van Gerven, M.A.J. Deep Neural Networks Reveal a Gradient in the Complexity of Neural Representations across the Ventral Stream. *J. Neurosci.* **2015**, *35*, 10005–10014. [CrossRef]
35. Naselaris, T.; Olman, C.A.; Stansbury, D.E.; Ugurbil, K.; Gallant, J.L. A voxel-wise encoding model for early visual areas decodes mental images of remembered scenes. *NeuroImage* **2015**, *105*, 215–228. [CrossRef]
36. Koide-Majima, N.; Nishimoto, S.; Majima, K. Mental image reconstruction from human brain activity: Neural decoding of mental imagery via deep neural network-based Bayesian estimation. *Neural Netw.* **2024**, *170*, 349–363. [CrossRef]
37. Shen, G.; Horikawa, T.; Majima, K.; Kamitani, Y. Deep image reconstruction from human brain activity. *PLoS Comput. Biol.* **2019**, *15*, e1006633. [CrossRef]
38. Kosslyn, S.M.; Thompson, W.L. When is early visual cortex activated during visual mental imagery? *Psychol. Bull.* **2003**, *129*, 723–746. [CrossRef]
39. Dijkstra, N.; Bosch, S.E.; van Gerven, M.A.J. Vividness of Visual Imagery Depends on the Neural Overlap with Perception in Visual Areas. *J. Neurosci.* **2017**, *37*, 1367–1373. [CrossRef]
40. Dijkstra, N.; Fleming, S.M. Subjective signal strength distinguishes reality from imagination. *Nat. Commun.* **2023**, *14*, 1627. [CrossRef]
41. Bartolomeo, P. The relationship between visual perception and visual mental imagery: A reappraisal of the neuropsychological evidence. *Cortex* **2002**, *38*, 357–378. [CrossRef]

42. Pearson, J. Reply to: Assessing the causal role of early visual areas in visual mental imagery. *Nat. Rev. Neurosci.* **2020**, *21*, 517–518. [CrossRef]
43. Cabbai, G.; Racey, C.; Simner, J.; Dance, C.; Ward, J.; Forster, S. Sensory representations in primary visual cortex are not sufficient for subjective imagery. *bioRxiv* **2024**, *in preprint*. [CrossRef]
44. Meng, M.; Chang, S.; Zhang, X.; Pearson, J. Imageless imagery in aphantasia: Decoding non-sensory imagery in aphantasia. 2023, *in preprint*. [CrossRef]
45. Weber, S.; Christophe, T.B.; Gorgen, K.; Soch, J.; Haynes, J.D. Working memory and imagery in early visual cortex. *Hum. Brain Mapp.* **2024**, *45*, e26590. [CrossRef]
46. Brown, R.; Lau, H.; LeDoux, J.E. Understanding the Higher-Order Approach to Consciousness. *Trends Cogn. Sci.* **2019**, *23*, 754–768. [CrossRef]
47. Fleming, S.M. Awareness as inference in a higher-order state space. *Neurosci. Conscious.* **2020**, *2020*, niz020. [CrossRef]
48. Lau, H.; Rosenthal, D. Empirical support for higher-order theories of conscious awareness. *Trends Cogn. Sci.* **2011**, *15*, 365–373. [CrossRef]
49. Liu, J.; Bayle, D.J.; Spagna, A.; Sitt, J.D.; Bourgeois, A.; Lehongre, K.; Fernandez-Vidal, S.; Adam, C.; Lambrecq, V.; Navarro, V.; et al. Fronto-parietal networks shape human conscious report through attention gain and reorienting. *Commun. Biol.* **2023**, *6*, 730. [CrossRef]
50. Iamshchinina, P.; Kaiser, D.; Yakupov, R.; Haenelt, D.; Sciarra, A.; Mattern, H.; Luesebrink, F.; Duezel, E.; Speck, O.; Weiskopf, N.; et al. Perceived and mentally rotated contents are differentially represented in cortical depth of V1. *Commun. Biol.* **2024**, *4*, 1069. [CrossRef]
51. Harrison, S.A.; Tong, F. Decoding reveals the contents of visual working memory in early visual areas. *Nature* **2009**, *458*, 632–635. [CrossRef]
52. Keogh, R.; Pearson, J. Attention driven phantom vision: Measuring the sensory strength of attentional templates and their relation to visual mental imagery and aphantasia. *Philos. Trans. R. Soc. B Biol. Sci.* **2021**, *376*, 20190688. [CrossRef]
53. Pounder, Z.; Jacob, J.; Jacobs, C.; Loveday, C.; Towell, T.; Silvanto, J. Mental rotation performance in aphantasia. *J. Vis.* **2018**, *18*, 1123. [CrossRef]
54. Nanay, B. Unconscious mental imagery. *Philos. Trans. R. Soc. London. Ser. B Biol. Sci.* **2021**, *376*, 20190689. [CrossRef] [PubMed]
55. Barry, D.N.; Barnes, G.R.; Clark, I.A.; Maguire, E.A. The neural dynamics of novel scene imagery. *J. Neurosci.* **2019**, *39*, 4375–4386. [CrossRef]
56. Lawrence, S.J.D.; van Mourik, T.; Kok, P.; Koopmans, P.J.; Norris, D.G.; de Lange, F.P. Laminar Organization of Working Memory Signals in Human Visual Cortex. *Curr. Biol. CB* **2018**, *28*, 3435–3440. [CrossRef] [PubMed]
57. Bergmann, J.; Petro, L.S.; Abbatecola, C.; Li, M.S.; Morgan, A.T.; Muckli, L. Cortical depth profiles in primary visual cortex for illusory and imaginary experiences. *Nat. Commun.* **2024**, *15*, 1002. [CrossRef]
58. Bastos, A.M.; Usrey, W.M.; Adams, R.A.; Mangun, G.R.; Fries, P.; Friston, K.J. Canonical Microcircuits for Predictive Coding. *Neuron* **2012**, *76*, 695–711. [CrossRef] [PubMed]
59. Bastos, A.M.; Vezoli, J.; Bosman, C.A.; Schoffelen, J.-M.; Oostenveld, R.; Dowdall, J.R.; De Weerd, P.; Kennedy, H.; Fries, P. Visual Areas Exert Feedforward and Feedback Influences through Distinct Frequency Channels. *Neuron* **2014**, *85*, 390–401. [CrossRef] [PubMed]
60. Keller, A.J.; Roth, M.M.; Scanziani, M. Feedback generates a second receptive field in neurons of the visual cortex. *Nature* **2020**, *582*, 545–549. [CrossRef]
61. Semedo, J.D.; Jasper, A.I.; Zandvakili, A.; Krishna, A.; Aschner, A.; Machens, C.K.; Kohn, A.; Yu, B.M. Feedforward and feedback interactions between visual cortical areas use different population activity patterns. *Nat. Commun.* **2022**, *13*, 1099. [CrossRef]
62. Wilming, N.; Murphy, P.R.; Meyniel, F.; Donner, T.H. Large-scale dynamics of perceptual decision information across human cortex. *Nat. Commun.* **2020**, *11*, 5109. [CrossRef]

vision MDPI

Article

In the Eyes of the Future: Eye Movement during Near and Distant Future Thinking

Mohamad El Haj [1,2,*] **and Ahmed A. Moustafa** [3,4]

1 CHU Nantes, Clinical Gerontology Department, Bd. Jacques Monod, 44093 Nantes, France
2 LPPL—Laboratoire de Psychologie des Pays de la Loire, Faculté de Psychologie, Université de Nantes, Chemin de la Censive du Tertre, BP 81227, CEDEX 3, 44312 Nantes, France
3 School of Psychology, Faculty of Society and Design, Bond University, Gold Coast, QLD 4226, Australia
4 Department of Human Anatomy and Physiology, The Faculty of Health Sciences, University of Johannesburg, Johannesburg 2006, South Africa
* Correspondence: mohamad.elhaj@univ-nantes.fr

Abstract: Research has suggested that near future events are typically viewed from a first-person (an own-eyes, also known as field) perspective while distant future events are typically viewed from a third-person (an observer) perspective. We investigated whether these distinct mental perspectives would be accompanied by distinct eye movement activities. We invited participants to imagine near and distant future events while their eye movements (i.e., scan path) were recorded by eye-tracking glasses. Analysis demonstrated fewer but longer fixations for near future thinking than for distant future thinking. Analysis also demonstrated more "field" mental visual perspective responses for near than for distant future thinking. The long fixations during near future thinking may mirror a mental visual exploration involving processing of a more complex visual representation compared with distant future thinking. By demonstrating how near future thinking triggers both "field" responses and long fixations, our study demonstrates how the temporality of future thinking triggers both distinct mental imagery and eye movement patterns.

Keywords: eye movement; future thinking; imagery; mental imagery; temporality

1. Introduction

In daily life, individuals dedicate significant time to envisioning the future. This future-oriented cognition aids in decision-making by offering a mental platform for evaluating, adjusting, and optimizing decisions, thereby enhancing problem-solving abilities and action planning [1–4]. Alternatively, future thinking may simply provide the ability to cope with the difficulties of daily life [5]. Beyond its practical functions, future thinking also serves an identity function, fostering a sense of self-continuity [6–8].

Regarding its temporality, future thinking allows simulating near and distant events. As we emphasize below, research has demonstrated that a key difference between near and distant future thinking is the mental visual perspective. More specifically, near distant future thinking typically implies an "own eyes visual perspective" while distant future thinking typically implies an "outside visual perspective". Building on this research, we investigate, for the first time, differences in eye movement patterns as associated with the construction of near and distant future events. By doing so, we hope to understand whether mental visual perspectives during different future thinking timescales may be accompanied by specific eye movement patterns.

Several studies have investigated mental visual perspectives for near vs. distant future thinking. These studies were inspired by the assumption that, when imagining ourselves across time, we can use a first-person (i.e., field) perspective involving looking through one's own eyes, or a third-person (i.e., observer) perspective involving looking at oneself from the outside [9]. The relationship between these two visual perspectives and future thinking was evaluated by D'Argembeau and Van der Linden [10], who invited participants to imagine an event that might happen in the near future (i.e., in the next year,) and another event that might happen in the distant future (i.e., in five to ten years). In each event, participants were invited to report whether they mentally "saw" the scene from their own perspective (i.e., field perspective) or "saw" themselves in their representation (i.e., observer perspective). Results demonstrated that near future events triggered more field perspective responses compared to distant events. Similar results were reported by Macrae et al. [11], who investigated whether perceivers use a different mental perspective when imagining their selves in the near and distant future. Results demonstrated that, whereas near future self-images were dominated by a first-person perspective, distant self-images were dominated by a third-person perspective. In a similar vein, Hamilton and Cole [12] invited participants to imagine an ideal-self that they would like to be in the near or distant future. Results demonstrated that the third-person perspective increased with future temporal distance. Similar findings were reported by Aikman et al. [13], who invited participants to imagine near or distant future events. The authors found that near future events were mainly associated with a first-person perspective. Together, research has consistently reported that, compared to their distant counterparts, near future events are typically viewed from a first-person perspective. Building on this research, we investigated whether different eye movement activities can be observed between distant and near future thinking.

In addition to drawing from previous research on mental visual perspective during future thinking, our study was inspired by investigations into eye movements during future thinking. El Haj and Lenoble [14] conducted a study where participants were asked to imagine future events and recall past events, while their eye movements (scan path) were recorded using an eye-tracker. Results revealed that future thinking elicited fewer fixations and saccade counts compared to past thinking. The authors attributed these fixations and saccades, observed during both past and future thinking, to mental imagery and the visual system's attempt to locate (via saccades) and activate (via fixations) stored mental representations. While the study of El Haj and Lenoble [14] was the first to investigate eye movement during future thinking, it did not examine the temporal distance (near vs. distant) or visual perspective (first- vs. third-person perspective) of these events.

The link between eye movement and mental imagery has been explored in studies investigating eye movement during past reminiscence. In a groundbreaking investigation, El Haj et al. [15] tasked participants with recalling past events and, in a control scenario, with performing a counting task aloud, while tracking their eye movement patterns with an eye-tracker. Their findings revealed a reduction in the number of fixations but an increase in the quantity, length, and duration of saccades during past reminiscence compared to the counting task. El Haj, Delerue, Omigie, Antoine, Nandrino, and Boucart [15] attributed eye movement activity during past thinking to visual imagery as required to generate the visual scene of the retrieved events. A similar assertion was posited by another study, linking the duration of fixations and saccades during the retrieval of emotional memories to the vividness of these recollections [16]. This connection between eye movement and mental imagery, as illuminated by research on eye movements during past and future reminiscence, can be grounded in the notion that autobiographical memories manifest in the form of visual mental images [17]. This proposition finds support in studies demonstrating superior autobiographical recall in individuals with strong visual imagery compared to those with weaker imagery [18], as well as neuropsychological investigations revealing autobiographical memory impairments in patients with lesions in vision-related brain regions, such as the right occipital lobe at the temporo-occipital junction [19].

To summarize, while previous research has investigated eye movement during future thinking [14], this prior research did not investigate the temporal distance (i.e., near vs. distant) nor the visual perspective (i.e., first- vs. third-person perspective) of these events. This comparison is worth consideration because research has consistently demonstrated that construction of near and distant events involves distinct mental visual perspectives [10–12].

Building on this prior research, we investigated whether the first-person perspective, as may be expected for near future thinking, may be accompanied by distinct eye movement pattern (e.g., long fixations). If this is the case, our study will demonstrate for the first time how the visual perspective of future thinking can be accompanied by distinct eye movement activities.

2. Method

2.1. Participants

The sample consisted of forty-five graduate/undergraduate, French-speaking students at the University of Nantes (25 females, M age = 21.09 years, SD = 3.73, M education = 13.24 years, SD = 4.54). We evaluated verbal episodic memory of the participants with the test of Grober and Buschke [20], with the maximum score being 16 points. The mean score of the participants was 12.16 (SD = 2.21). Additionally, working memory was evaluated through span tasks, where participants repeated strings of numbers in both forward and backward order. The mean score of the participants was 7.31 (SD = 1.92) on the forward span and 5.31 (SD = 1.28) on the backward span. Based on these memory assessments, we excluded four participants who performed two standard deviations below the norms on the Grober and Buschke [20] test, as well as one participant who scored three points below the norm on the backward span task. Furthermore, four participants with a history of psychiatric/neurological disorders and seven participants experiencing difficulties with eye movement recording were excluded. After applying these criteria, the final sample comprised 45 participants. This sample size was predetermined using G*Power [21] [within-subject measurement *t*-tests as our experimental design involved two repeated conditions (i.e., near vs. distant future), 95% power, an estimated probability of making a Type I error as 0.05, and a medium effect size of 0.25 [22]]. In the final sample, no significant differences were observed regarding gender [X^2 (1, N = 45) = 0.55, p = 0.45]. The study was validated by the ethic committee of the University of Nantes (reference IORG0011023).

2.2. Procedures

The procedures are illustrated in Figure 1.

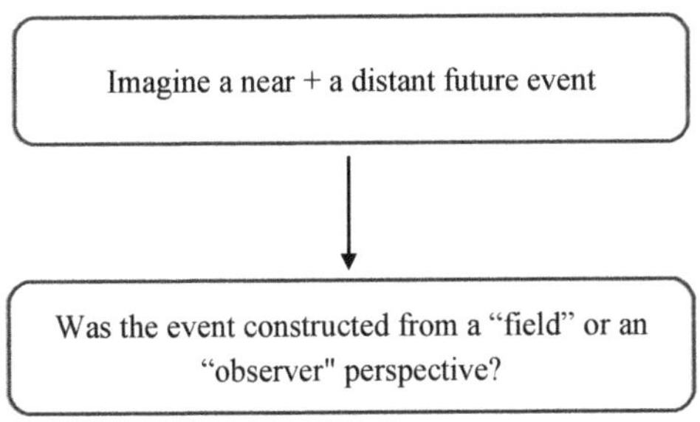

Figure 1. Illustration of procedures.

2.2.1. Future Thinking and Eye Movement Recording

Participants were instructed to verbally envision two future events: one occurring in the near future and the other in the distant future. While narrating these events, participants wore eye-tracking glasses and faced a blank white wall. They were prompted to be specific about the details of each event, including where it would take place, who would be involved, and their emotional experiences. Following the protocol outlined by D’Argembeau and Van der Linden [10], participants were given two minutes to describe each event, with the duration clearly communicated to aid in structuring their narratives. The order of the near and distant future trials was counterbalanced across participants. Participants wore eye-tracking glasses (Pupil Lab) with a gaze position accuracy of $<0.1°$ and a 200 Hz sampling rate. These glasses emit near-infrared light onto the eyes, detecting and tracking pupil and corneal reflection and, thus, providing precise measurements of gaze direction and fixation points. Eye movement data were recorded and processed using the Pupil Capture software V 3.5.1. Calibration of the eye-tracking glasses was conducted before each trial by instructing participants to focus on a black cross positioned at the center of the wall. The experiment took place in a quiet room at the psychology department of the University of Nantes, with consistent lighting conditions and participants seated approximately 30 to 50 cm away from the wall. They were instructed not to look beyond the wall but were free to explore its entire surface. Unlike previous research utilizing screen-based designs, our study employed a wall-based setup to allow for unrestricted visual exploration, minimizing constraints on the visual system.

2.2.2. Mental Visual Perspective

We replicated the procedures of previous research using the "field/observer" paradigm [23,24], as originally described by Nigro and Neisser [9]. Following the near and distant future thinking tasks, participants were briefed on the concept of visual perspective. Specifically, they were informed that when imagining an event, they might perceive it from either a field or an observer perspective. We clarified that the field perspective entails visualizing the event from a first-person viewpoint, as if experiencing it through their own eyes. Alternatively, the observer perspective involves visualizing the event from an external viewpoint, where they may see themselves within the scene, along with their surroundings. Participants were then prompted to verbally indicate their chosen "field/observer" perspective for each imagined future event.

2.2.3. Dependent Variables

In alignment with prior studies [14,25], we analyzed several eye movement parameters, including fixation count, fixation duration, saccade count, saccade duration, and total saccade amplitude. Fixation and saccade counts were normalized to per-minute rates due to variations in the duration of future thinking tasks among participants. Fixation and saccade durations represented the mean duration of fixations and saccades, respectively, measured in milliseconds. Total saccade amplitude denoted the overall angle covered by saccades during the task. Additionally, blinks were automatically excluded from the dataset, and gazes exceeding a horizontal deviation of 2° were excluded (comprising 4.8% of the dataset). To assess reconstruction time consistently between conditions, we measured the total duration of recording. Furthermore, we tallied the frequency of "field" and "observer" responses provided by participants for both near and distant future thinking tasks.

2.2.4. Statistical Analysis

Regarding eye movement (scores are provided in Table 1), we compared each dependent variable (i.e., number of fixations, fixation duration, number of saccades, saccade duration, total amplitude, and duration of recording) between near and distant future thinking using the paired samples *t*-test. Regarding mental visual perspective, we compared the number of "field" and "observer" responses (provided in Table 2) between near and distant future thinking with chi-square tests. We provided effect sizes by using Cohen's d [22]: 0.20 = small, 0.50 = medium, 0.80 = large. Level of significance was set as $p \leq 0.05$.

Table 1. Characteristics of eye movements during near and distant future thinking.

	Near Future	Distant Future
Fixation count per min	57.38 (17.23) *** (less fixations in near than in distant future thinking)	79.11 (29.23)
Fixation duration in ms	794.36 (510.69) ** (longer fixations in near than in distant future thinking)	564.41 (216.64)
Saccade count per min	62.69 (28.10) ns	66.82 (23.26)
Saccade duration in ms	55.70 (27.34) ns	53.36 (19.16)
Total amplitude of saccades	1622.02 (857.69) ns	1425.45 (630.14)
Duration of recording in msec	82,604.34 (21,032.63) ns	83,311.29 (25,162.89)

Note. Standard deviations are given between brackets; maximum duration of recording was 120,000 ms; differences were significant at ** $p < 0.01$, *** $p < 0.001$; ns differences were not significant.

Table 2. Number of "field" and "observer" responses during near and distant future thinking.

	Near Future	Distant Future
Field	32 *	15
Observer	13 *	30

Note. For each condition (near and distant future thinking), the total number of "field" and "observer" responses was 45 as participants ($n = 45$) were required to provide either an "field" or an "observer" response; differences were significant at * $p < 0.05$.

2.3. Results

2.3.1. Fewer, but Longer, Fixations for Near Than for Distant Future

Compared with distant future thinking, near future thinking triggered fewer fixations [$t(44) = 3.94$, $p < 0.001$, Cohen's $d = 0.73$] but longer fixation duration [$t(44) = 2.77$, $p = 0.008$, Cohen's $d = 0.56$]. No significant differences were observed between near and distant future thinking regarding number of saccades [$t(44) = 0.87$, $p = 0.39$, Cohen's $d = 0.16$], duration of saccades [$t(44) = 0.44$, $p = 0.66$, Cohen's $d = 0.08$], or amplitude of saccades [$t(44) = 1.15$, $p = 0.25$, Cohen's $d = 0.22$]. No significant differences were observed between near and distant future thinking regarding duration of recording [$t(44) = 0.36$, $p = 0.72$, Cohen's $d = 0.08$].

2.3.2. More "Field" Perspective Responses for Near Than for Distant Future

Compared with distant future thinking, near future thinking triggered more "field" [X^2 (1, $N = 47$) = 6.15, $p = 0.013$, Cohen's $d = 0.78$] and fewer "observer" responses [X^2 (1, $N = 47$) = 6.49, $p = 0.011$, Cohen's $d = 0.80$]. Near future thinking triggered more "field" than "observer" responses [X^2 (1, $N = 45$) = 8.02, $p = 0.004$, Cohen's $d = 0.93$], whereas distant future thinking triggered fewer "field" than "observer" responses [X^2 (1, $N = 45$) = 6.15, $p = 0.01$, Cohen's $d = 0.79$].

3. Discussion

We investigated eye movement and mental visual perspective during near and distant future thinking. Analysis demonstrated fewer but longer fixations for near than for distant future thinking. Analysis also demonstrated more "field" mental visual perspective responses for near than for distant future thinking. These findings demonstrate, for the first time, how the temporality of future thinking triggers distinct eye movement activities.

To begin with the eye movement activities as observed in our study, near future thinking triggered fewer fixations and longer fixation duration during near than during distant future thinking. In other words, when constructing near future events, participants demonstrated fewer, but longer, fixations than when constructing distant future events. This pattern of fixations can be attributed to the complexity of mental visual processing during near and distant future thinking. Generally speaking, the duration of fixations reflects the allocation of attention to a given stimuli [26] and stimulus complexity [27]. Furthermore, the duration of fixations reflects information processing [28]. Therefore, the long fixations during near future thinking may mirror a mental visual exploration involving processing of a more complex visual representation compared with distant future thinking.

We suggest that the long fixations during near future thinking can mirror the processing of complex visual representation. Our assumption can be supported by the Construal-Level Theory [29–31]. According to this theory, the construction of near events (e.g., travel next month) typically triggers a concrete and context-rich representation (e.g., booking a convenient hotel) compared to distant events, which typically imply a decontextualized representation that conveys only the gist or general meaning of the event (e.g., enjoying next year's holidays). In other words, as future events become temporally distant, representations become more abstract, probably because people typically possess limited information about the distant future events. Supporting this model, research has demonstrated that, compared with the near future, distant future self-representations are typically more abstract [32]. For example, distant future self-representations typically involve less complexity and a greater degree of schematicity compared with near future self-representations [33,34]. By demonstrating how near future thinking involves a concrete and context-rich representation, the Construal-Level Theory supports our attempt to associate the long fixations during the construction of near future events to the complex visual representation of these events.

The discovery of fewer but longer fixations during near future thinking holds significant implications for understanding cognitive processes in future thinking. Our findings suggest that temporally proximal events prompt individuals to engage in more elaborate mental simulations, involving detailed contextual information and sensory-rich representations. Moreover, the identification of distinct eye movement patterns corresponding to different temporal distances provides insights into the temporal dynamics of mental imagery and perspective-taking processes. These findings may have practical applications in various domains, including education, marketing, and clinical psychology, by informing interventions aimed at promoting forward planning, decision-making, and goal-setting behaviors. Moreover, the consequences of our findings extend beyond theoretical insights into future thinking. By elucidating the relationship between temporal distance and eye movement patterns, our study opens avenues for practical applications in various domains. For instance, in education, understanding how individuals mentally simulate near and distant future events can inform instructional strategies aimed at enhancing long-term planning and goal-setting skills. In marketing, insights into the cognitive processes underlying future thinking can guide the development of persuasive messaging and advertising campaigns that resonate with consumers' temporal perspectives. Additionally, in clinical psychology, knowledge of how temporal distance influences mental imagery can inform therapeutic interventions targeting future-oriented cognition in individuals with mood disorders or anxiety. Overall, our findings not only advance our understanding of future thinking but also offer practical implications for promoting adaptive decision-making and behavior across diverse contexts.

Besides being associated with the complexity of visual representation, the long fixations during near future thinking can be associated with the first-person perspective. As demonstrated in our study, most near future events triggered a "field" mental visual perspective, replicating previous research on the prevalence of this perspective during near future thinking [10–12]. The "field" perspective can be associated with the vividness of the visual representation of near future events. Generally speaking, this perspective involves imagining the event as it being experienced in the reality [9]. Furthermore, this perspective, as typically observed during near future thinking, involves the construction of vivid mental images [35]. Therefore, and because the "field" perspective has been associated with the vividness of the retrieved mental image, it is tempting to consider the long fixations, as observed during near future thinking, as a behavioral indicator of both the "field" perspective and its vividness.

Although it is tempting to propose that the first-person perspective, as observed during near future thinking, can be expressed in terms of long fixations, this suggestion should be considered with some caution. While our study demonstrates that near future thinking triggers both long fixations and "field" responses, we provide no evidence that the eye movement activities during future thinking (i.e., the few and long fixations during near future thinking, or even the non-significant differences between near and distant future thinking in terms of saccades) are the result or origin of mental visual processing. As mentioned in the introduction here (see above), eye movement in general has been considered an index of visual imagery during memory retrieval [15,16,36,37]. Research has also attributed eye movement during future thinking to the attempt by the visual system to find (through saccades) and activate (through fixations) stored mental representations [14]. While prior research, as well as our current study, provides a new behavioral insight, in its current state of development, this research only demonstrates how past and future thinking trigger both specific mental visual perspective and specific eye movement patterns (e.g., near future thinking triggers both "field" responses and long fixations, as demonstrated in our study). That being said, this research provides an exciting and novel departure from the traditional research widely assessing mental visual imagery during both past and future thinking in terms of subjective experience, by adding methods to measure eye movements.

Traditional research on mental visual imagery during both past and future thinking has demonstrated not only how the mental visual perspective (i.e., first- vs. third- person) can be influenced by temporality (i.e., near vs. distant events) but also how this perspective can be influenced by emotion. For instance, trauma memories have been widely associated with a third-person perspective [38], probably because this perspective involves avoidance and is less affect-provoking [39]. Further, the first-person perspective has been considered a predictor of the severity of post-traumatic stress disorder symptoms up to one year after the traumatic event, probably because this avoidance strategy plays a role in the maintenance of trauma [40]. The same thing can be said for depression, as the retrieval of negative memories from a third-person perspective has been linked to maladaptive avoidant strategies [41,42]. Therefore, it would be of interest to evaluate whether the third-person perspective during memory retrieval in post-traumatic stress disorder and depression can be associated with a specific pattern of eye movement activities. It would be also of interest to investigate this issue for future thinking because post-traumatic stress disorder, depression, and, critically, anxiety, have been associated with taking a third-person perspective during future thinking [43]. In our view, the assessment of eye movement during past and future thinking in these clinical disorders will provide a valuable behavioral indicator of cognitive processing in patient populations.

One limitation of our study is the assessment of only one near and one distant future event. Future research can replicate our study using a large number of near and distant future scenarios. As suggested previously, future research should also replicate our study on clinical disorders (e.g., post-traumatic stress disorder, depression, and anxiety) as these disorders have been associated with deficits in visual perspective. Furthermore, while our study utilized the field/observer paradigm to assess visual imagery, it would have been intriguing to incorporate direct manipulation of visual imagery. For example, inviting participants to construct concrete versus abstract conceptions of the future could have provided valuable insights into eye movement during future thinking.

It would be valuable to situate our findings within the context of recent advancements in computer vision research. For instance, Yang, et al. [44] conducted a thorough review of human parsing, which involves partitioning humans in images or videos into semantic parts. Their study sheds light on the challenges and potential research directions in this field, providing insights into how individuals mentally parse and integrate visual information during cognitive tasks such as future thinking. Additionally, Singh, et al. [45] introduced Latent Graph Attention, a framework for efficiently incorporating a global context into image processing tasks. This advancement aligns with our investigation into how individuals construct mental representations of future events, as it facilitates the integration of contextual information across spatially distant points in images. By aligning our findings with these developments in computer vision, future research stands to benefit from a multidisciplinary perspective that enhances our understanding of the cognitive processes underlying future thinking and mental imagery. Moreover, future replications could leverage recent advancements in machine vision optics. Innovations such as light-trapping-structure vertical Ge photodetectors enable high-speed and high-sensitivity detection of optical signals, thereby improving the precision of eye movement analysis during future thinking [46]. Additionally, advancements in optical computing architectures and adaptive multiscale imaging mechanisms offer opportunities for real-time analysis of eye movement data with enhanced accuracy and efficiency [47]. Moreover, the integration of on-chip metasurfaces in augmented reality holography presents new avenues for immersive and multifunctional augmented reality displays [48], potentially enhancing gaze-based interactions and personalized content delivery. Incorporating insights from these advancements into eye-tracking research could lead to significant improvements in understanding eye movement patterns during future thinking tasks.

To summarize, research on the mental visual perspective during future thinking has mainly investigated this perspective in terms of subjective measures. Our current study provides an exciting and novel departure from this research by proving new behavioral insight into the mental visual perspective during future thinking. While research on eye movement and past and future thinking in general stills in its infancy in terms of available empirical studies or even established models, we believe that this generation of research will yield new insights into the relationship between eye movement and cognitive characteristics of both past and future thinking.

Author Contributions: Conceptualization, M.E.H.; methodology, M.E.H. validation, A.A.M.; investigation, M.E.H.; resources, A.A.M.; writing—original draft preparation, M.E.H.; writing—review and editing, A.A.M. All authors have read and agreed to the published version of the manuscript.

Funding: This research received no external funding.

Institutional Review Board Statement: The study was validated by the ethic committee of the University of Nantes (reference IORG0011023, 10 November 2022).

Informed Consent Statement: Informed consent was obtained from all subjects involved in the study.

Data Availability Statement: Raw data is available upon request to the correspondent author.

Conflicts of Interest: The authors have no conflicts of interest to declare.

References

1. Suddendorf, T.; Busby, J. Making decisions with the future in mind: Developmental and comparative identification of mental time travel. *Learn. Motiv.* **2005**, *36*, 110–125. [CrossRef]
2. D'Argembeau, A.; Renaud, O.; Van der Linden, M. Frequency, characteristics and functions of future-oriented thoughts in daily life. *Appl. Cogn. Psychol.* **2011**, *25*, 96–103. [CrossRef]
3. Stawarczyk, D.; D'Argembeau, A. Neural correlates of personal goal processing during episodic future thinking and mind-wandering: An ALE meta-analysis. *Hum. Brain Mapp.* **2015**, *36*, 2928–2947. [CrossRef] [PubMed]
4. Berntsen, D.; Bohn, A. Remembering and forecasting: The relation between autobiographical memory and episodic future thinking. *Mem. Cogn.* **2010**, *38*, 265–278. [CrossRef] [PubMed]
5. Killingsworth, M.A.; Gilbert, D.T. A Wandering Mind Is an Unhappy Mind. *Science* **2010**, *330*, 932. [CrossRef] [PubMed]
6. D'Argembeau, A.; Lardi, C.; Van der Linden, M. Self-defining future projections: Exploring the identity function of thinking about the future. *Memory* **2012**, *20*, 110–120. [CrossRef] [PubMed]
7. Rathbone, C.J.; Conway, M.A.; Moulin, C.J. Remembering and imagining: The role of the self. *Conscious. Cogn.* **2011**, *20*, 1175–1182. [CrossRef] [PubMed]
8. Schacter, D.L.; Benoit, R.G.; Szpunar, K.K. Episodic Future Thinking: Mechanisms and Functions. *Curr. Opin. Behav. Sci.* **2017**, *17*, 41–50. [CrossRef] [PubMed]
9. Nigro, G.; Neisser, U. Point of view in personal memories. *Cogn. Psychol.* **1983**, *15*, 467–482. [CrossRef]
10. D'Argembeau, A.; Van der Linden, M. Phenomenal characteristics associated with projecting oneself back into the past and forward into the future: Influence of valence and temporal distance. *Conscious. Cogn.* **2004**, *13*, 844–858. [CrossRef]
11. Macrae, C.N.; Mitchell, J.P.; Tait, K.A.; McNamara, D.L.; Golubickis, M.; Topalidis, P.P.; Christian, B.M. Turning I into me: Imagining your future self. *Conscious. Cogn.* **2015**, *37*, 207–213. [CrossRef]
12. Hamilton, J.; Cole, S.N. Imagining possible selves across time: Characteristics of self-images and episodic thoughts. *Conscious. Cogn.* **2017**, *52*, 9–20. [CrossRef] [PubMed]
13. Aikman, S.N.; Doyle-Portillo, S.; Bell, C.R.; Simmons, N. When I saw me standing there: First-person and third-person memories and future projections, and how they relate to the self AU—Verhaeghen, Paul. *J. Cogn. Psychol.* **2018**, *30*, 438–452. [CrossRef]
14. El Haj, M.; Lenoble, Q. Eying the future: Eye movement in past and future thinking. *Cortex* **2018**, *105*, 97–103. [CrossRef] [PubMed]
15. El Haj, M.; Delerue, C.; Omigie, D.; Antoine, P.; Nandrino, J.L.; Boucart, M. Autobiographical recall triggers visual exploration. *J. Eye Mov. Res.* **2014**, *7*, 1–7. [CrossRef]
16. El Haj, M.; Nandrino, J.L.; Antoine, P.; Boucart, M.; Lenoble, Q. Eye movement during retrieval of emotional autobiographical memories. *Acta Psychol.* **2017**, *174*, 54–58. [CrossRef]
17. Conway, M.A. Episodic memories. *Neuropsychologia* **2009**, *47*, 2305–2313. [CrossRef] [PubMed]
18. Vannucci, M.; Pelagatti, C.; Chiorri, C.; Mazzoni, G. Visual object imagery and autobiographical memory: Object Imagers are better at remembering their personal past. *Memory* **2016**, *24*, 455–470. [CrossRef] [PubMed]
19. Greenberg, D.L.; Eacott, M.J.; Brechin, D.; Rubin, D.C. Visual memory loss and autobiographical amnesia: A case study. *Neuropsychologia* **2005**, *43*, 1493–1502. [CrossRef]
20. Grober, E.; Buschke, H. Genuine memory deficits in dementia. *Dev. Neuropsychol.* **1987**, *3*, 13–36. [CrossRef]

21. Faul, F.; Erdfelder, E.; Lang, A.G.; Buchner, A. G*Power 3: A flexible statistical power analysis program for the social, behavioral, and biomedical sciences. *Behav. Res. Methods* **2007**, *39*, 175–191. [CrossRef] [PubMed]
22. Cohen, J. Statistical Power Analysis. *Curr. Dir. Psychol. Sci.* **1992**, *1*, 98–101. [CrossRef]
23. Libby, L.K.; Eibach, R.P.; Gilovich, T. Here's Looking at Me: The Effect of Memory Perspective on Assessments of Personal Change. *J. Personal. Soc. Psychol.* **2005**, *88*, 50–62. [CrossRef] [PubMed]
24. Libby, L.K.; Eibach, R.P. Looking back in time: Self-concept change affects visual perspective in autobiographical memory. *J. Pers. Soc. Psychol.* **2002**, *82*, 167–179. [CrossRef]
25. Rice, H.J.; Rubin, D.C. Remembering from any angle: The flexibility of visual perspective during retrieval. *Conscious. Cogn.* **2011**, *20*, 568–577. [CrossRef] [PubMed]
26. Sperling, G.; Weichselgartner, E. Episodic theory of the dynamics of spatial attention. *Psychol. Rev.* **1995**, *102*, 503–532. [CrossRef]
27. McConkie, G.W. Eye movements and perception during reading. In *Eye Movements in Reading: Perceptual and Language Processes*; Rayner, K., Ed.; Academic Press: New York, NY, USA, 1983; pp. 65–96.
28. Zelinsky, G.J. Understanding scene understanding. *Front. Psychol.* **2013**, *4*, 954. [CrossRef] [PubMed]
29. Trope, Y.; Liberman, N. Temporal construal. *Psychol. Rev.* **2003**, *110*, 403–421. [CrossRef] [PubMed]
30. Trope, Y.; Liberman, N. Temporal construal and time-dependent changes in preference. *J. Pers. Soc. Psychol.* **2000**, *79*, 876–889. [CrossRef]
31. Trope, Y.; Liberman, N. Construal-level theory of psychological distance. *Psychol. Rev.* **2010**, *117*, 440–463. [CrossRef]
32. Nussbaum, S.; Liberman, N.; Trope, Y. Predicting the near and distant future. *J. Exp. Psychol. Gen.* **2006**, *135*, 152–161. [CrossRef] [PubMed]
33. Wakslak, C.J.; Nussbaum, S.; Liberman, N.; Trope, Y. Representations of the self in the near and distant future. *J. Pers. Soc. Psychol.* **2008**, *95*, 757–773. [CrossRef] [PubMed]
34. Heller, D.; Stephan, E.; Kifer, Y.; Sedikides, C. What will I be? The role of temporal perspective in predictions of affect, traits, and self-narratives. *J. Exp. Soc. Psychol.* **2011**, *47*, 610–615. [CrossRef]
35. La Corte, V.; Piolino, P. On the Role of Personal Semantic Memory and Temporal Distance in Episodic Future Thinking: The TEDIFT Model. *Front. Hum. Neurosci.* **2016**, *10*, 385. [CrossRef]
36. Lenoble, Q.; Janssen, S.M.J.; El Haj, M. Don't stare, unless you don't want to remember: Maintaining fixation compromises autobiographical memory retrieval. *Memory* **2019**, *27*, 231–238. [CrossRef] [PubMed]
37. El Haj, M. When you look at your past: Eye movement during autobiographical retrieval. *Conscious. Cogn.* **2024**, *118*, 103652. [CrossRef] [PubMed]
38. Berntsen, D.; Willert, M.; Rubin, D.C. Splintered memories or vivid landmarks? Qualities and organization of traumatic memories with and without PTSD. *Appl. Cogn. Psychol.* **2003**, *17*, 675–693. [CrossRef]
39. McIsaac, H.K.; Eich, E. Vantage point in traumatic memory. *Psychol. Sci.* **2004**, *15*, 248–253. [CrossRef]
40. Kenny, L.M.; Bryant, R.A.; Silove, D.; Creamer, M.; O'Donnell, M.; McFarlane, A.C. Distant memories: A prospective study of vantage point of trauma memories. *Psychol. Sci.* **2009**, *20*, 1049–1052. [CrossRef]
41. Williams, J.M.; Scott, J. Autobiographical memory in depression. *Psychol. Med.* **1988**, *18*, 689–695. [CrossRef]
42. Williams, J.M. Capture and rumination, functional avoidance, and executive control (CaRFAX): Three processes that underlie overgeneral memory. *Cogn. Emot.* **2006**, *20*, 548–568. [CrossRef] [PubMed]
43. Coles, M.E.; Turk, C.L.; Heimberg, R.G.; Fresco, D.M. Effects of varying levels of anxiety within social situations: Relationship to memory perspective and attributions in social phobia. *Behav. Res. Ther.* **2001**, *39*, 651–665. [CrossRef] [PubMed]
44. Yang, L.; Jia, W.; Li, S.; Song, Q. Deep Learning Technique for Human Parsing: A Survey and Outlook. *Int. J. Comput. Vis.* **2024**. [CrossRef]
45. Singh, A.; Bhambhu, Y.; Buckchash, H.; Gupta, D.K.; Prasad, D.K. Latent Graph Attention for Enhanced Spatial Context. *arXiv* **2023**, arXiv:2307.04149.
46. Chen, D.; Zhang, H.; Liu, M.; Hu, X.; Zhang, Y.; Wu, D.; Zhou, P.; Chang, S.; Wang, L.; Xiao, X. 67 GHz light-trapping-structure germanium photodetector supporting 240 Gb/s PAM-4 transmission. *Photon. Res.* **2022**, *10*, 2165–2171. [CrossRef]
47. Ma, G.; Yu, J.; Zhu, R.; Zhou, C. *Optical multi-imaging*–casting accelerator for fully parallel universal convolution computing. *Photon. Res.* **2023**, *11*, 299–312. [CrossRef]
48. Shi, Y.; Wan, C.; Dai, C.; Wang, Z.; Wan, S.; Zheng, G.; Zhang, S.; Li, Z. Augmented Reality Enabled by On-Chip Meta-Holography Multiplexing. *Laser Photonics Rev.* **2022**, *16*, 2100638. [CrossRef]

vision

MDPI

Review

A Systematic Review of Aphantasia: Concept, Measurement, Neural Basis, and Theory Development

Feiyang Jin [1,2], Shen-Mou Hsu [3] and Yu Li [1,4,*]

1 Applied Psychology Program, Department of Life Sciences, BNU-HKBU United International College, Zhuhai 519087, China; sofiajin0830@gmail.com
2 Department of Applied Social Sciences, The Hong Kong Polytechnic University, Hong Kong, China
3 Imaging Center for Integrated Body, Mind and Culture Research, National Taiwan University, Taipei 10617, Taiwan; smhsu@ntu.edu.tw
4 Guangdong Provincial Key Laboratory of Interdisciplinary Research and Application for Data Science, BNU-HKBU United International College, Zhuhai 519087, China
* Correspondence: yuli@uic.edu.cn

Abstract: People with aphantasia exhibit the inability to voluntarily generate or form mental imagery in their minds. Since the term "aphantasia" was proposed to describe this, it has gained increasing attention from psychiatrists, neuroscientists, and clinicians. Previous studies have mainly focused on the definition, prevalence, and measurement of aphantasia, its impacts on individuals' cognitive and emotional processing, and theoretical frameworks synthesizing existing findings, which have contributed greatly to our understanding of aphantasia. However, there are still some debates regarding the conclusions derived from existing research and the theories that were constructed from various sources of evidence. Building upon existing endeavors, this systematic review emphasizes that future research is much needed to refine the definition and diagnosis of aphantasia, strengthen empirical investigations at behavioral and neural levels, and, more importantly, develop or update theories. These multiple lines of efforts could lead to a deeper understanding of aphantasia and further guide researchers in future research directions.

Keywords: aphantasia; visual imagery; mental imagery; vividness; cognitive functioning; mental health

1. Introduction

In 1880, Galton first systematically elucidated and studied individual differences in visual imagery, designing specific questions to investigate related phenomena [1]. Visual imagery ability refers to the capacity of individuals to create mental images of objects or scenes that are not in front of their eyes [2], such as closing one's eyes and imagining a brightly colored apple. Visual imagery has been shown to be closely associated with various cognitive functions and everyday activities. The active generation of visual imagery is an important cognitive ability of humans. However, the human capacity for visual imagery can be considered a spectrum, with some individuals exhibiting exceptionally vivid visual imagery, which is referred to as hyperphantasia [3,4], while other individuals can only produce limited or even deficient visual imagery. In 2015, Zeman and colleagues introduced the term "aphantasia" to describe absent or marked reduced mental imagery in individuals [5].

Presently, investigations into aphantasia concentrate on delineating its nature, assessing its frequency, devising measurement methods, examining its influence on cognitive and emotional functions, identifying related disorders, probing its neural underpinnings, and developing conceptual models. In particular, studies examining cognitive functions primarily explore how aphantasia impacts memory, atemporal and future imagination, spatial imagery, mental rotation abilities, and visual search abilities (e.g., [6,7]). Neuroimaging studies indicate a possible link between the lack of visual imagery and the visual cortex (e.g., [2,7,8]).

Aphantasia is still a relatively new topic in psychological research and has not received widespread attention from researchers. This review aims to summarize the existing empirical studies and theoretical viewpoints on aphantasia. First, our discussion will concentrate on the definition, prevalence, heritability, and assessment of aphantasia. Second, we will synthesize the findings regarding aphantasia's influence on cognitive functions. Third, we will examine its potential associations with related disorders. Fourth, we will summarize existing evidence on the neural basis of aphantasia. Fifth, we will critically appraise the principal theoretical models. Last, we provide future research prospects for aphantasia, aiming to further guide researchers in possible research directions.

2. Literature Retrieval and Screening

This study undertook a systematic search of English-language articles published between 2015 and the present (Figure 1). The commencement year of 2015 was selected due to its significance as the year in which Zeman et al. first coined the term "aphantasia" to characterize the absence and marked reduction of visual imagery. The databases used include Scopus, PubMed, ProQuest, and Web of Science. All the relevant journal articles, master's and doctoral theses, conference papers, and preprints were included in the search. As of 14 August 2024, an initial retrieval yielded 255 literature sources, from which 113 duplicates were removed. Other screening methods and a careful examination of the references cited in pertinent articles added 28 more literature sources to the pool. An in-depth review of titles, abstracts, and full texts for these 170 literature sources led to the exclusion of 10 inaccessible papers, 76 papers not centrally concerned with aphantasia, and 19 non-empirical studies, finally resulting in a curated collection of 65 pertinent literature sources.

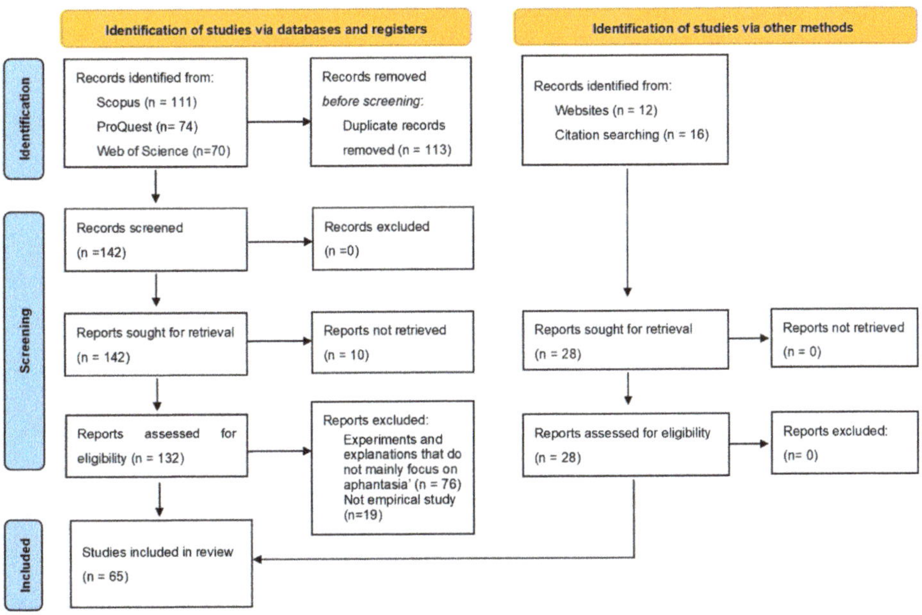

Figure 1. Flow diagram illustrating the research searching and screening.

We evaluated the quality of these studies and classified their evidence. A study quality evaluation was conducted based on the criteria used in previous works [9,10]. The details of the criteria for the quality assessment are presented in Table S1 (Supplementary Materials).

Out of the 65 studies included, 58 studies were cross-sectional or case-control studies. Seven studies were case reports or ecological studies. The overall quality scores (%) were categorized as follows: excellent quality (score ≥81%), good quality (between 61 and 80%), fair quality (between 41 and 60%), poor quality (between 21 and 40%), and very poor quality (≤20%). The evaluation results showed that 58 studies were assessed as "excellent" (N = 22) or "good" (N = 36), 8 studies were assessed as "fair", and only one study was identified as of "poor" quality (Table S2 in Supplementary Materials).

Based on these 65 papers, this review provides a detailed analysis, evaluation, and summary of the existing empirical research on aphantasia. It should be emphasized that the key studies published before 2015, though not using the term "aphantasia", have also been considered in discussions if necessary. Reviews and commentaries are further incorporated into this study. In this review, we used "aphantasia" to refer to an absence of mental imagery in any modality for consistency, although the majority of the reviewed studies focused on visual imagery deficits. The detailed content is summarized as follows (see Tables S3–S15 in Supplementary Materials for the details of each study included).

3. Definition, Measurement, and Prevalence of Aphantasia

Aphantasia is conceptually defined as the inability to generate mental imagery. However, variations exist in the details of the conceptual definitions [11,12]. Blomkvist reviewed various definitions of aphantasia [4–7,13–19] and identified three primary areas of disagreement: whether aphantasics exhibit impairments solely in visual imagery, whether a distinction is made between the absence of voluntary and involuntary imagery, and whether aphantasia primarily concerns the generation of imagery [11]. Regarding the first aspect, aphantasia has also been associated with deficits in non-visual modalities (see Section 4.1). Some studies do not emphasize "visual imagery" when describing its absence, instead using the broader term "mental imagery" to refer to various modalities [7,15,18]. It is recommended that the consistent term "aphantasia" be used and that the modality be specified when necessary, such as "auditory aphantasia" [20]. Concerning voluntariness, some researchers explicitly defined aphantasia as the lack of voluntary visual imagery [6,7,17], while others did not specifically address this aspect [14,16]. Krempel and Monzel call for simply using a broader characterization of aphantasia in the field as there is no solid evidence of preserved involuntary imagery in aphantasics [21]. Regarding the third aspect, increasing evidence suggests that aphantasia is linked to other deficits in cognitive and emotional processes (see Section 4).

In measuring visual imagery deficits, most research utilizes scores from the Vividness of Visual Imagery Questionnaire (VVIQ, [22]) as the primary identification and diagnostic tool [12]. The VVIQ comprises 16 questions and employs a 5-point Likert scale. Participants are asked to rate their visual imagery ability on a scale ranging from "No imagery at all, you only 'know' that you are thinking of the object" to "Perfectly clear and as vivid as real seeing", to differentiate between individuals with visual imagery deficits, those with normal imagery abilities, and those with excessive visual imagery [22]. According to the criteria, a score of 16 denotes a complete absence of visual imagery, with a score of 17–32 indicating vague/dim imagery. In recent years, a new reliable method for identifying visual imagery deficits has emerged, known as the binocular rivalry task [13,23]. Prior instructions to imagine one of the images increase the likelihood of perceiving that image during the subsequent binocular rivalry task. Participants with visual imagery deficits exhibit minimal evidence of image-based binocular rivalry priming [13,23]. The VVIQ-2, an extended version of the VVIQ containing 32 items [24], and a variant of the VVIQ have also been used in the field [4]. Recent research has also found that pupillary responses to light can indicate visual imagery deficits, as affected individuals lack an imagery pupillary light response but still exhibit a perceptual pupillary light response [25]. Additionally, rhythmic visual flickering typically induces illusory percepts, and compared to control groups, individuals with visual imagery deficits are less likely to experience complex and vivid illusions [26,27]. These findings further highlight the physiological differences between individuals with visual imagery deficits and control groups, offering an objective measure for assessing visual imagery intensity.

The vividness of the image can be viewed as a continuous characteristic forming a normal distribution curve with aphantasia occupying the left tail and hyperphantasia occupying the right tail. Before the term "aphantasia" was introduced, the most cited work based on a single question about visual imagery indicated a prevalence rate of 2.1–2.7% for a total absence of visual imagery in the general population [28]. In the field, the VVIQ was widely used for measuring the vividness of visual imagery. Different operational definitions have led to varying prevalence rates. Zeman et al. found that the prevalence rate of aphantasia, where imagery is completely absent, was 0.7% (VVIQ = 16), and that of moderate aphantasia with a VVIQ score of 16–23 was 2.6% [4]. Using a VVIQ score of 16–32 covering "totally absent" to "vague/dim" imagery ratings, Dance et al. found a prevalence rate of 3.9% across two separate samples [29]. When using a score of 16, the rate was only 0.8%. Later, Monzel, Vetterlein, and Reuter conducted meta-analyses and found a prevalence rate of 4.8% across all the included studies and 3.5% in studies using the VVIQ [30]. Recently, Takahashi et al. found in a large sample of Japanese individuals that the rate was 0.07% when using a score of 16 and 3.6% when using a score of 17–32 [31]. A study with a large sample of Brazilian university students reported a prevalence rate of 5.9% based on VVIQ scores of 16–32 [32]. However, another study noted a self-reported prevalence rate of 8.9% in the general adult population, but not all self-reported aphantasics showed low imagery scores on the VVIQ, resulting in only 1.5% [33]. Overall, the prevalence rates of aphantasia are inconsistent due to the use of different criteria across studies. If the strictest criterion is used, the rate would be extremely low. It is important to use consistent standards when discussing the prevalence rates.

For the effects of gender, studies have not found any gender differences in the prevalence of aphantasia [29,34]. Although previous studies found that women tend to score higher on object imagery measures compared to men [35,36], this difference is not statistically significant enough to conclude that men are more likely to be aphantasics. Age has also been considered as a factor influencing aphantasia. However, there is no consensus on how age affects aphantasia [34,37]. Additionally, whether the prevalence of aphantasia varies across different professions remains controversial [4,37,38], despite Galton's initial suggestion that scientists might have weaker visual imagery abilities as a group [1]. This controversy could be explained by Blajenkova et al.'s opinion that scientists may not lack all types of visual imagery but may be specifically deficient in object imagery [39,40]. Similarly, visual artists tend to excel in spatial imagery rather than object imagery [40].

Aphantasia can be either congenital or acquired [41,42]. It is more common in individuals with a family history of lacking visual imagery, suggesting a genetic basis. However, a recent study found no significant genetic association, indicating the need for further research to explore the genetic evidence of aphantasia [43]. Acquired aphantasia can provide insights into the causes and mechanisms of aphantasia. This form of aphantasia can be triggered by various events, including craniocerebral injuries, emotional disorders, stroke, and postoperative complications [41,42,44]. A case study reported a female who developed aphantasia after contracting COVID-19 [45]. Psychological and psychiatric factors should be carefully considered in the assessment of aphantasia [46,47] as mental illnesses such as depression and anxiety can significantly impact the vividness of visual imagery [46].

4. Aphantasia and Cognitive Processing

4.1. *Visual and Non-Visual Imagery Ability*

Visual imagery ability is associated with other forms of imagery, such as olfactory and auditory imagery. In a pioneering study by Zeman et al., 10 out of 21 individuals with visual imagery deficits reported impairments in all forms of imagery, including auditory, olfactory, gustatory, tactile, motor, and bodily imagery, to some extent [5]. A certain number of aphantasics reported a lack of all forms of mental imagery, while some only have deficits in visual imagery [31]. In a large-scale study, 54.2% of aphantasics with visual deficits reported deficits in all types of imagery [4]. Similar findings have been confirmed in other studies, revealing that some aphantasics with visual deficits also reported deficits in other forms of mental imagery (34%, [48]; 26%, [6]). Takahashi and Gyoba reported a person with aphantasia who exhibited a complete deficit in various types of images (e.g., visual, olfactory, pain, tactile, gustatory, and somatic images) in vividness and a substantial deficit in auditory images [49]. Furthermore, individuals with hyperphantasia, who exhibit exceptionally vivid visual imagery, tend to experience more vivid imagery in other types of imagery as well [4]. There may be overlapping mechanisms underlying visual and auditory imagery, as many aphantasia participants report weak or absent auditory imagery [50–52] and individuals who lack auditory imagery, which is termed "anauralia", often show visual imagery deficits [51]. However, some aphantasics self-reported higher auditory imagery than the mean score of the control group [53]. Using cluster analysis, Dawes et al. recently found that aphantasia is heterogeneous and has two subtypes: visual aphantasia, which selectively shows an absence of visual imagery, and multisensory aphantasia, which shows an inability to generate any sensory modality of mental imagery [54].

Overall, some individuals with aphantasia may also experience deficits in all types of mental imagery. However, there is still a lack of large-scale studies to establish a reliable and consistent picture, as well as plausible explanations of the neural mechanisms for understanding these phenomena. It is worth noting that most of the previous measurements of visual and other forms of mental imagery in research have relied on self-report questionnaires (e.g., the VVIQ, the Auditory Imagery Questionnaire, and the Questionnaire upon Mental Imagery). The latest research revealed no significant difference between aphantasics and the control group in an auditory imagery task, even though most of the aphantasic participants reported an impairment in auditory imagery [53]. This suggests that, although self-report questionnaires are convenient for large samples, there may be a meta-cognitive effect in the data. Participants may not actually lack genuine mental imagery but rather report deficits based on cognitive differences and misunderstandings of the questionnaire items. More precise, reliable, and objective measurement methods could help clarify the relationships between different forms of mental imagery deficits by mitigating the risks of participants' meta-cognitive influences. Alternatively, when using self-report questionnaires, participants should be provided with clear explanations and clarifications of the items to minimize misunderstandings about the content of the questionnaires.

4.2. Aphantasia and Memory

Existing research on the impact of visual imagery impairment on memory has mainly focused on working memory, episodic memory, autobiographical memory, and object and spatial memory. Overall, aphantasics do not exhibit deficits in simple working memory tasks [55–58]. However, it has been reported that aphantasics perform worse than control groups in tasks requiring fine-grained visual working memory, which are cognitively more demanding and require participants to remember smaller differences [55,56,58]. Jacobs et al. reported an aphantasia individual who performed worse than the controls only on the most difficult visual working memory trials requiring a high level of precision [15]. The difficulty level appears to be a crucial factor in visual working memory tasks. In terms of episodic memory, individuals with aphantasia demonstrated significantly lower performance than the control group [6,59,60]. Similarly, aphantasia individuals have been reported to experience difficulties in autobiographical memory [4–6,16,55,60–62]. Regarding spatial memory, which involves the processing of an object's location information, two recent studies did not find any deficits in aphantasics [6,17]. For object memory, aphantasics showed poorer performance in object memory in a drawing task requiring object processing [17]. Siena and Simons used both subjective and objective measures of object and spatial memory and found that aphantasia participants showed no objective deficits in memory performance, indicating that some aphantasics might show a deficit in the awareness of mental imagery [63].

It should be noted that new measurements of memory performance have been used in existing research, deviating from traditional scales and experimental tasks. For instance, drawing has been used to complement the assessment of memory performance in participants [17]. In the case of drawing, aphantasics and control groups were instructed to draw scene images from real-world memory. It was found that aphantasics remembered fewer objects and used fewer colors to draw, and relied more on verbal scaffolding to compensate for the lack of visual imagery [17].

Overall, memory is a relatively well-researched aspect in the field of aphantasia. Many studies have employed questionnaires and experiments to measure participants' memory performance. In visual working memory, aphantasics exhibit similar performance to control groups, which could be explained by their potential use of semantic encoding or other representational strategies to aid them in completing working memory tasks [7]. However, some studies indicate that aphantasics perform worse in these visual working memory tasks [15]. The inconsistency of these findings may be attributed to task characteristics, measurement methods, and individual differences among participants. For instance, the heterogeneous outcome included only one individual with aphantasia, suggesting that the observed differences in working memory could be influenced by sample size or the unique characteristics of the case study [15]. As for autobiographical memory, all the results seem to converge on the conclusion that individuals with aphantasia have poorer autobiographical memory. Differences in object memory and spatial memory performance between the aphantasics group and the control group may be related to their abilities in object imagery and spatial imagery. These differences may be explained through the neural mechanisms of the ventral and dorsal pathways of visual processing, which will be discussed more comprehensively later.

4.3. Aphantasia and Object and Spatial Imagery

Object imagery refers to visualizing the image appearance of objects and scenes, such as their shape, color, and brightness, while spatial imagery refers to visualizing the spatial relationships and movements of objects and their components, and spatial transformation [64]. Existing research on the performance of individuals with aphantasia in object and spatial imagery abilities has not reached a consensus and can be categorized into three types of studies. The first type of study indicates that aphantasia individuals score lower in object imagery abilities but show no significant differences from control groups in spatial imagery performance [6,17,52,60]. The second type of study shows that aphantasics score significantly lower in object imagery than control groups but perform better in spatial imagery tests [13,55]. In the last category of results, individuals with aphantasia demonstrate poorer performance in both object and spatial imagery abilities [65,66]. Based on these findings and a review of the literature, researchers have proposed two types of visual imagery impairments: object aphantasia and spatial aphantasia [39,67].

The finding that object imagery is impaired while spatial imagery remains unaffected can be also explained by the functional division of visual pathways. Aphantasics' ventral pathway for object processing may be impaired, while the dorsal pathway for spatial processing remains intact ([7,68], also see below). To better explain spatial aphantasia, researchers have proposed that this type of aphantasia may be related to functional changes in the visual dorsal stream projecting to the frontal lobe [67]. However, this explanation still lacks further support from neuroimaging studies.

4.4. Aphantasia and Atemporal and Future Imagination

Atemporal and future imagination are related to voluntarily imagining general events and future events, respectively. Autobiographical interviews on atemporal and future imagination show that aphantasia participants have poorer abilities in atemporal and future imagination, with fewer details in their imagined scenarios [16,60]. Evidence suggests that atemporal and future imagination abilities are related to autobiographical memory [16]. This may help explain why aphantasics show poorer abilities in atemporal and future imagination, as their autobiographical memory abilities are also lower compared to control groups.

4.5. Aphantasia and Mental Rotation Task Performance

The mental rotation task has been widely used to investigate individuals' spatial abilities [69]. It has been reported that, compared to control groups, individuals with aphantasia showed equal [16,53,70] or even higher accuracy in mental rotation tasks [55,65,66,71]. However, aphantasics also required longer response times in these tasks [58,65,71].

Aphantasia and control groups performed similarly in mental rotation tasks, possibly due to the use of different strategies to solve the tasks [66,71]. Pounder and his colleagues suggested that aphantasics may employ non-visual processes in these tasks [58]. For example, the patient in Furman et al.'s study self-reported using a spatial and kinesthetic strategy instead of low-level visual object imagery in the mental rotation task [13,66]. Specifically, he used a first-person path-following strategy (i.e., spatial strategy) containing kinesthetic features, using his body as a reference. This hypothesis is supported by the performance of congenitally or early blind individuals in mental rotation tasks, as they did not show differences from control groups in completing complex spatial tasks, indicating that mental rotation tasks can be completed independently of visual processes [72]. Kay and his colleagues suggested that aphantasics are more likely to use analytic strategies, which do not depend on features and orientation [71]. Additionally, the mental rotation task is considered to rely on an individual's spatial imagery ability, and many individuals with aphantasia report having similar or even better spatial imagery abilities compared to control groups [13,60].

4.6. Aphantasia and Visual Searching Ability

Visual search is a commonly used skill in everyday life that helps us find the things we need, even if the desired target is within our current field of vision. When performing visual search tasks, aphantasics exhibit significantly slower speeds compared to control groups [73,74]. This difference could be explained by the varying involvement of visual imagery in top-down strategies between the two groups [74,75]. However, in the Moriya's task, which examines participants' attentional guidance in visual search tasks ([76]; participants are first asked to imagine a color primed by a color word such as "blue" and then to indicate whether one of the two colored squares presented on the left and right sides has an opening at the top or bottom), the priming process seems to be similar between the aphantasia group and the control group [73]. This can be explained by the characteristics of the task, where the instructions in the Moriya's task are overly complex, resulting in a lack of visual priming and only non-visual priming [73]. These findings suggest that visual imagery could affect the way people perceive the world. Future research would benefit from designing materials that are more ecologically valid, such as incorporating environmental cues into the experimental design.

5. Aphantasia and Disorders and Emotional Processing

5.1. Emotion

Mood may influence the vividness of mental imagery [4]. By contrast, mental imagery was also found to affect individuals' emotion experience. Due to absent or reduced visual imagery, aphantasics experience reduced emotional engagement and less sympathy for characters in stories [77] and exhibit lower levels of fear response when reading scary materials [78]. Mental imagery ability was found to be positively associated with empathy elicited by descriptions, but not associated with empathy elicited by pictures [79]. Additionally, aphantasics seem to experience lower emotional intensity when listening to music [80]. This may be because vivid imagery makes thoughts more realistic, triggering stronger responses in the brain's emotional circuitry and thereby amplifying emotions [81]. Future research can continue to explore different types of emotions and incorporate neuroimaging methods to validate previous findings.

5.2. Mental Health

Although aphantasia may impact individuals' cognition and emotional processing and potentially cause some personal distress, it is not currently included in any common clinical diagnostic systems. Furthermore, the impact of aphantasia on individuals' daily lives and personal distress is relatively minimal and not sufficient to classify it as a psychological disorder [30,82]. However, research has found associations between aphantasia and certain conditions or diseases [83]. Moreover, scientists have proposed that progressive aphantasia may be a precursor to dementia [84].

Existing research has focused on the impact of aphantasia on mental health, such as depression, anxiety, distress, and well-being. Research has revealed that the aphantasia group showed no significant differences compared to the control group in terms of depression, anxiety, and state-trait anxiety [16,81]. However, Monzel, Vetterlein, and Reuter found that 34.7% of aphantasia participants reported distress caused by the lack of visual imagery [30]. Nevertheless, the researchers also suggested that these negative emotions were relatively weak and did not significantly affect individuals' daily lives [30].

5.3. Post-Traumatic Stress Disorder (PTSD)

Previous research has established a link between mental imagery and PTSD, a condition characterized by re-experiencing traumatic events through unwanted and recurring intrusive memories and nightmares [85,86]. Aphantasics have been found to experience fewer intrusions and exhibit less avoidance behaviors following trauma [6], both of which are predictors of PTSD. Additionally, the aforementioned lack of visual imagery and its association with reduced emotional intensity may also contribute to their lower likelihood of developing PTSD. However, it is important to note that researchers emphasize the limited protective effect of visual imagery deficits when individuals with such deficits face stressful events in their daily lives [6].

5.4. Autism

Individuals with aphantasia frequently score higher on the Autism Quotient Questionnaire (AQ) and are more likely to be classified within the autism spectrum [16,87]. This association may be related to the connection between imagery and mentalizing. Compared to control groups, aphantasics exhibit slightly reduced atemporal and future imagination, as well as episodic memory [6,16,60], which are also characteristics of autistic spectrum disorder [88,89]. Additionally, limited or absent visual imagery can impair the theory of mind ability in aphantasics, leading to difficulties in social skills [87]. These factors may explain why individuals with aphantasia tend to have higher scores on the AQ, indicating a potential connection between visual imagery deficits and autistic spectrum disorder.

5.5. Prosopagnosia

Aphantasia has been associated with reduced facial recognition ability [31]. In a recent study, 5.9% of participants in the spatial aphantasia group were identified as having prosopagnosia [67], and face recognition difficulties were found to be more common in individuals with aphantasia [16,90]. Aphantasics also reported difficulties with facial recognition tasks [41]. Additionally, aphantasics have been associated with lower confidence in tasks involving facial perception [91]. In a study of developmental prosopagnosia, about 20% of participants with developmental prosopagnosia reported a comorbidity with aphantasia [92]. However, aphantasia does not significantly impact accuracy in tasks involving the construction of facial composites and face recognition tasks [93], as individuals with aphantasia can still create facial composites from memory similarly to control groups [90]. Nevertheless, some researchers have argued that the scores of aphantasics in tests are not sufficient to diagnose them with prosopagnosia [66]. Overall, individuals with aphantasia tend to exhibit poorer face recognition abilities. However, this may be influenced by the scales used in experiments, as these scales may elicit meta-cognitive effects [16]. In other words, individuals with aphantasia may have reduced confidence in their own facial recognition abilities due to their lack of vivid facial imagery, which can manifest as relatively lower scores on self-report measures.

6. The Neural Basis of Aphantasia

Although numerous studies have focused on the neural basis of mental imagery [7,8,94,95], there are limited studies utilizing modern neuroimaging techniques such as magnetic resonance imaging (MRI) and electroencephalography (EEG) to explore the neural basis of visual imagery deficits. Using resting-state and task-based MRI, Milton et al. recently compared the brain activities of three groups: aphantasics, typical controls, and hyperphantasics [16]. The results showed that individuals with hyperphantasia exhibit stronger connectivity between the prefrontal regions and visual-occipital network compared to the aphantasia group. When comparing the visualization and perception of famous faces and places, individuals with hyperphantasia and the control group exhibited greater frontal and parietal activation compared to the aphantasia group [16]. Another functional MRI study showed that the aphantasic group exhibited decreased activation in the hippocampus and increased activation in the visual-perceptual cortex during an autobiographical memory task [96]. However, the control group displayed strong negative task-based functional connectivity between the hippocampus and the visual cortex during the task, and the resting-state functional connectivity between these two areas could predict visualization skills [96]. Furthermore, using transcranial magnetic stimulation to induce changes in brain activity, Dupont, Papaxanthis, Madden-Lombardi and Lebonet found that there was no increase in the amplitude of motor-evoked potentials triggered in the target right index finger in the aphantasia group, indicating a lack of corticospinal excitability in individuals with aphantasia during motor simulation [97]. Similar findings were also observed in an action reading task, which involved motor simulation [98].

Case studies are also a crucial source of neural evidence. Lesion studies have found that patients showed intact visual imagery after brain lesions restricted to the occipital cortex [99–102], suggesting that early visual areas are not involved in visual imagery. Moreover, patients with damage to the anterior part of the temporal lobe, particularly in the left hemisphere, often report an inability to generate visual images [103–105], which is consistent with observations of a left hemisphere bias [106]. Zeman et al. reported a patient who showed reduced activation in the occipitotemporal regions during an imagery task [107]. In a recent study, a patient with an absence of visual imagery ability showed selective lesions in a specific area of the left fusiform gyrus and a portion of the right lingual gyrus, demonstrating a causal role of the left fusiform gyrus in visual imagery [41]. Therefore, the fusiform region "might act as a neural interface between sensory information coming from the occipital cortex and semantic processing in the anterior temporal lobe" during visual perception, and "could endow semantic memories with visual information during visual imagery" ([100], p. 517). An EEG study with source reconstruction reported that during a visual imagery task, an aphantasic begins the evoking phase from the left temporal area while lacking activation of the occipital and parietal lobes, which are associated with visual image vividness [66].

Due to the limited evidence in these studies, it is challenging to draw a consistent picture of the aberrant brain activity in aphantasia. However, we can still draw inspiration from neuroimaging studies based on individual differences in mental imagery. Spagna et al. recently conducted a meta-analysis of fMRI studies of visual imagery and found that visual imagery recruits several fronto-parietal areas and a specialized area in the left fusiform gyrus [94]. The specialized area was labeled the fusiform imagery node (FIN), referring to a brain network node specifically responsible for voluntary visual mental imagery [94,108,109]. Liu et al. found that imagery tasks activated the left frontal–parietal regions, the FIN, and areas in the ventral temporal cortex, which were similarly activated in the aphantasia and control groups [110]. However, the connectivity between the FIN and the frontoparietal regions is reduced in aphantasics [110]. Together, these brain-lesion cases and fMRI studies demonstrate that the fusiform gyrus is a core area for visual imagery, and damage to or impairment of it could lead to deficits in visual imagery. Additionally, abnormalities in the fusiform gyrus have been associated with prosopagnosia, which may explain the facial recognition deficits observed in individuals with visual imagery disorders [111,112].

A related debate concerns the involvement of the early visual cortex in (impaired) visual imagery [100,113]. Using stimulation, Kosslyn et al. found that the early visual cortex is causally involved in visual imagery [114]. Bergmann et al. documented that a smaller V1 size is associated with stronger but less precise imagery, indicating an anatomical basis [115]. Keogh et al. suggested a causal relationship between cortical excitability in the early visual cortex and the intensity of visual imagery [116]. However, Meng et al. suggested that an imagery-related representation exists in the primary visual cortex of aphantasics, despite the absence of visual imagery, though the representation contains less or transformed sensory information [117]. The results from Cabbai et al. showed a dissociation between V1 representations and subjective imagery [118]. Bartolomeo et al. [100] argued that the left fusiform gyrus plays a crucial role in visual imagery, rather than the early visual cortex, and that the involvement of the early visual cortex [114] might be modulated by downstream areas. Dijkstra recently reviewed existing evidence and proposed that "imagery can recruit the early visual cortex, but that does not mean that it always does so" [95].

7. Theory Development

Individuals with imagery deficits can still have imagination capabilities similar to those without such deficits, such as creating novels and movies. Based on this, Arcangeli attempts to distinguish between mental imagery and sensory imagination [119]. Mental imagery can be considered a type of mental content (e.g., the appearance of an apple). In contrast, sensory imagination is a special psychological attitude that involves the recreation of perceptual experiences. According to this theory, most individuals previously defined as having aphantasia in past research may actually have a deficiency in sensory imagination rather than in mental imagery, which could explain why individuals with visual imagery deficits can still perform imaginative tasks.

Nanay explains aphantasia from the perspective of both conscious and unconscious visual imagery [18]. Previous research has found that individuals with aphantasia do not exhibit the imagery priming effect in binocular rivalry tasks; yet some participants can generate vivid dreams and perform visual imagery tasks similarly to control groups. To explain this phenomenon, Nanay distinguishes between conscious and unconscious visual imagery abilities, both of which could be generated voluntarily or involuntarily [18]. One type of aphantasia involves a fundamental lack of visual imagery, while the other type involves the ability to generate visual imagery without conscious awareness [73]. However, this theory is challenged by Blomkvist [11], who argues that it does not fully explain the issues of episodic memory or imagination of atemporal and future events in visual imagery deficits, as it does not provide a link between mental imagery and the episodic processes in episodic memory and episodic imagination. Additionally, there is research suggesting that imagery tasks can be accomplished through cognitive strategies that do not rely on imagery [91,120–122]. A recent study using both implicit and explicit priming tasks found that participants with aphantasia did not show priming effects, suggesting that aphantasia is an inability to generate visual imageries rather than an impairment in conscious awareness of images [123].

To provide a better explanation for aphantasia, Blomkvist proposed enhancements to the Constructive Episodic Simulation Hypothesis (CESH) model [11]. The original CESH posits that memory and imagination involve three key processes: the semantic retrieval process, episodic retrieval process, and (re)combination process [124]. Blomkvist expanded on this model by adding three new components: "memory indices, differing episodic retrieval mechanisms for all kinds of sensory information, and spatial retrieval mechanisms" [11]. These additions aim to provide a new theoretical explanation for visual imagery deficits. According to this updated theory, the mechanism of the episodic system is deficient, resulting in the loss of visual imagery. However, this theory still lacks empirical research data to support it further.

The theoretical framework of the ventral and dorsal pathways in the visual system has been proposed for many years [125–127] (although there are ongoing debates about this framework, [128,129]. Pearson proposed that the ventral (or "what") pathway is associated with object information, while the dorsal (or "where") pathway is associated with location and spatial features [7]. The two aspects of visual imagery, object imagery and spatial imagery, are also likely generated through the ventral and dorsal pathways [39]. Damage to the ventral pathway could impair individuals' ability to visualize the appearance of objects, while damage to the dorsal pathway is associated with a disrupted ability in spatial imagery. These processes can be dissociated in aphantasia, which may explain why individuals with visual imagery deficits perform similarly to, or even better than, control groups in spatial imagery tasks [7,68]. Bergmann and Ortiz-Tudela also suggested that visual feedback pathways used during episodic and schematic memory retrieval may be different depending on the two visual processing streams: episodic memory retrieval involves both the "what" and "where" streams, while schematic memory retrieval primarily involves the "where" stream [130]. In other words, aphantasia may be associated with differences in the "what" stream.

Inspired by neural models of mental imagery [7,8] and existing sources of neural evidence, Zeman tried to depict candidate neural mechanisms of extreme imagery, aphantasia, and hyperphantasia [2]. In the brain mapping picture, there are five functional clusters, each involved in a unique role in mental imagery: the frontal cortex for initiating imagery generation, the parietal cortex for interacting with the frontal cortex to generate imagery during which attentional and spatial aspects of imagery are mediated, the temporal cortex, including limbic structures for enabling access to the semantic and episodic memories that determine what to visualize, and higher-order visual areas (e.g., the FIN) for visualizing imagery. The extent to which activity in the early visual cortex (e.g., the V1) is required for imagery is still debated. Moreover, the connections between these clusters are also essential for imagery generation. From this mapping picture, Zeman [2] proposes five candidates showing neural atypicality in extreme imageries: variations in the strengths of the top-down feedback connection between higher-order regions (the frontal cortex) serving as cognitive control and modality-specific areas (e.g., the visual cortex) activated by sensory imagery; variations in the structure and function of the frontal cortex involved in generating imagery; variations at the level of higher-order visual areas, including the FIN proposed by Spagna et al. [94]; variations in the anatomical structure of the early visual cortex, including the V1; and variations in the saliency involving the parietal areas of the frontoparietal control system. In this proposal, Zeman provided future directions for researchers in the field [2]. A neural model constructed and refined from accumulated evidence would add to a better understanding of aphantasia.

8. Summary and Future Directions

This review primarily examines the definition, prevalence, and measurement methods of aphantasia, as well as its impacts on individual cognitive and emotional processing, associated disorders, neural basis, and theory development. Research on aphantasia is continuously progressing, and its descriptions are becoming increasingly clear. However, several debated areas remain, and the study of its impacts on individuals and underlying mechanisms faces several challenges. Future investigations into aphantasia should focus on the following aspects.

8.1. Clarify Definition and Diagnosis

It is evident that the current research on aphantasia is still in its early infancy. One of the primary concerns is the definition of aphantasia. Clear definitions serve as the foundation for advancing further research and facilitating effective communication among researchers. The existing literature presents variations in the definition of aphantasia [12].

For example, some definitions do not explicitly distinguish between voluntary and involuntary imagery (see a review by [21]). The cut scores for the diagnosis of aphantasia vary across studies (see Section 3). The inconsistent use of the arbitrary cut scores is considered a hindrance to cross-study comparison and communication among researchers. Moreover, the use of cut scores may also discourage efforts to examine individual differences in imagery. Additionally, more effective, objective tools for non-clinical and clinical purposes should be developed [131,132] rather than relying predominantly on the VVIQ or its variants [133].

The terminology in the field has recently been debated [20,134–136]. While the debate continues on whether the term "aphantasia" specifically refers to the absence of visual mental imagery or encompasses a deficit of all types of mental imagery (see Section 3), most current research on aphantasia focuses on visual imagery. Other terms have been proposed to refer to an absence of other modalities of imagery. For auditory imagery, Hinwar and Lambert introduced the term "anauralia" to refer to an absence of auditory imagery and "hyperauralia" to refer to the experience of extremely vivid auditory imagery [51]. Dance, Ward, and Simner proposed the term "dysikonesia" to refer to multisensory or global aphantasia [48]. In this context, Monzel, Mitchell et al. believe that these new terms complicate the field, making communication less effective for researchers and the public, and therefore advocate for the use of the simple term "aphantasia," which is widely known [20]. When referring to an absence of a specific modality of imagery, it is easy to use modality-specific terms (e.g., "visual/auditory/multisensory aphantasia").

These divergencies across studies create difficulties in the classification of participants and scientific communications among researchers and the public, therefore hindering progress toward a better understanding of aphantasia. Researchers in the field should reach a consensus on these issues to facilitate further development.

8.2. Strengthen Behavioral Research

Aphantasia research is still at its early stages, and there is limited literature in several research directions. Future studies should build upon existing efforts to systematically examine this phenomenon to better understand aphantasia. At the behavioral level, more research is urgently needed within and beyond existing research directions. First of all, together with a clear and consistent definition of aphantasia, researchers should focus on examining the nature of aphantasia, including but not limited to investigating effects of demographic variables (e.g., age and gender) and their interactions in large datasets, developing new assessment or diagnostic tools with high reliability and validity, elucidating perceptual and cognitive processes required by tasks used to elicit mental imagery, and determining the comorbidity rate of two or more subtypes of aphantasia in large samples. Second, researchers should investigate how aphantasia is associated with other psychological aspects (e.g., conscious experience, [18,137–139], the influences of aphantasia on emotional processing and disorders (see Section 4), and possible mechanisms underlying these relationships (e.g., [140]). Third, research on hyperphantasia, the opposite extreme characterized by vivid mental imagery, can also help contribute to a better understanding of aphantasia. Viewing mental imagery ability as a continuum could be more realistic and scientifically useful in understanding the absence of or reduction in imagery in people with aphantasia. Fourth, researchers should use these different sources of evidence as a foundation to propose or create intervention programs for people with aphantasia. The effectiveness of these interventions can also help researchers better understand the nature of this phenomenon, fostering further advances.

8.3. Discover Neural Bases

Unraveling the neural bases of aphantasia has recently become a prominent research topic. Neuroimaging evidence has been instrumental in understanding the manifestations of aphantasia and in constructing neural models. Employing neuroimaging methods, future research can explore various variables to gather more neural evidence. Specifically, researchers can utilize the high spatial resolution of MRI techniques to investigate regional activation and inter-regional connections (structural and functional connectivity) involved in the absence or reduction of mental imagery. Previous functional MRI studies have identified areas involved in mental imagery, based on which neural models have also been proposed (e.g., [14,115,116]; see a review by [7]). Although some studies of aphantasia have been conducted [16,107,110], it is difficult to accumulate convergent evidence for localizing aberrant brain regions and connections. Using EEG and magnetoencephalography (MEG) techniques with high temporal resolution [141,142], researchers can examine the neural dynamics of the processes involving mental imagery in people with and without aphantasia (e.g., [66]). For example, Xie et al. found shared alpha-band neural representations in visual imagery and perception [143]. Combining spatial and temporal neural evidence across different studies, or even using fusion techniques (e.g., EEG-MRI fusion, [144]) can contribute to a better understanding of aphantasia by incorporating different types of information into a single picture, further facilitating the construction of neural models. Recent technical advances in multi-variate pattern analysis (e.g., [145,146]) may also be powerful in understanding the neural correlates of aphantasia, as demonstrated in previous studies of mental imagery [111,117,143]. Non-invasive brain stimulation techniques [147] can be powerful tools for discovering causal evidence of the involvement of brain areas in mental imagery. Together, the convergence of behavioral and neural evidence not only helps elucidate various aspects of aphantasia but is also useful in building behavioral and neural models that explain this phenomenon [2]. In addition, evidence from other sources of investigation should also be encouraged. It has been reported that aphantasia has a genetic basis [4] and that dopamine plays an important role in generating mental imagery [148].

8.4. Construct and Refine Theories

Good theories help humans systematically and scientifically understand their minds [149,150]. Although aphantasia has garnered increasing attention over the past decade, and has also gained interest in art, philosophical, and theoretical discussions [80,83,151–160], empirical studies examining this phenomenon remain limited. The field is still in its early stages. A lack of theories for synthesizing the existing literature is one of the field's significant characteristics. Although some accounts and models have been proposed to interpret behavioral findings [11,18,119], they have focused on certain aspects. Neural models for synthesizing neural evidence (e.g., [2]) are largely unknown or require further examination. One of the main reasons is that both behavioral and neural empirical studies are very limited in number and diversity (see above), resulting in insufficient evidence to be integrated into a theory or model.

In addition to proposing new theories or models, we advocate for efforts to use existing theories or models of mental imagery [7,8,11,109] as a foundation to theoretically guide future directions on researching aphantasia. Modifications to these models or theories will be made as more evidence is collected. We believe these changes will either refine existing theories or help foster new ones. For instance, theories that explain the mechanisms of visual imagery can be adapted to account for the absence of such imagery in individuals with aphantasia. By modifying these models to include the cognitive and neural underpinnings of aphantasia, researchers can create a more robust theoretical framework that encompasses both the presence and absence of mental imagery.

Given that aphantasia may extend beyond visual imagery to include other modalities, such as auditory or tactile imagery, theories should be developed to address these broader aspects. This could involve creating a unified model that categorizes different types of aphantasia (e.g., visual, auditory, multisensory) and explores how these modalities interact within cognitive processes. Such a model would not only clarify the nature of aphantasia but also facilitate comparisons with individuals who possess typical imagery capabilities.

8.5. Encourage Direct and Conceptual Replications

In recent years, concerns about replication in psychology have increased (e.g., [161,162]). Replicability is a crucial indicator of the reliability and stability of research findings [163]. Conducting replicable experiments to demonstrate the replicability of the findings lays a solid foundation for further investigations. There are two types of replications, each with different requirements. Conceptual replication deliberately modifies key elements of the original program to test the robustness of the phenomenon or the generalizability of the theoretical claims, while direct replication involves recreating the original experiment, which has been questioned regarding its impossibility of implementation [164].

The inconsistencies in research findings on aphantasia, which can undermine confidence in the literature, call for the consideration of replicability in existing studies, requiring numerous replication studies to confirm the reliability and stability of results. For example, there are limited studies on the relationship between aphantasia and atemporal and future imagination, as well as visual search. Although consistent conclusions have been reached, different experimental designs are still needed to validate the accuracy and replicability of the results (conceptual replications). Moreover, researchers have explored the relationships between attentional templates [19], sensory sensitivity [48], verbal overshadowing [120], mid-level characteristics of drawings [165], and aphantasia, but these are isolated studies lacking cross-validation from other research, indicating the necessity of conducting direct and conceptual replication studies.

To facilitate replication efforts, aphantasia researchers should adopt open science practices, such as sharing data, materials, and protocols. By making these more transparent, the scientific community of aphantasia can more easily conduct replications and build upon existing work. Open access to the datasets of research published or to be published can also encourage more collaboration and foster a collective effort to understand this phenomenon more deeply.

9. Conclusions

In this study, we have provided a comprehensive review of aphantasia, encompassing its definition, prevalence, measurement methods, empirical research, and relevant theories. The existing literature has contributed to a good understanding of aphantasia. The absence of visual imagery can have significant implications for an individual's cognitive and emotional processing and is also associated with certain psychological disorders. Studies on the neural bases of aphantasia are gradually increasing, especially in recent years. Following these endeavors, theories and neural models have been proposed to summarize existing behavioral and neural evidence. The research of visual imagery deficits is continually growing, although there are still areas of debate. However, a consensus regarding the theoretical framework has not yet been achieved. The theories still lack sufficient empirical support, leaving a significant gap. We have emphasized several points to which researchers can make contributions, such as clarifying diagnoses, strengthening behavioral and neuroimaging research, constructing and refining theories, and encouraging replications. New work on the different aspects of aphantasia is not only beneficial for expanding researchers' understanding of cognitive and emotional processes and their relationships with other disorders, but is also helpful for building theories or model.

Supplementary Materials: The following supporting information can be downloaded at: https://www.mdpi.com/article/10.3390/vision8030056/s1, Table S1. Criteria for quality assessment of the studies; Table S2. Scores and quality categories of the 65 studies evaluated; Table S3. Prevalence; Table S4. Performance on computer tasks; Table S5. Acquired aphantasia; Table S6. Non-visual imagery ability; Table S7. Memory; Table S8. Object and spatial imagery; Table S9. Atemporal and future imagination; Table S10. Mental rotation task performance; Table S11. Visual search ability; Table S12. Emotion; Table S13. Aphantasia and related disorders; Table S14. Neural Basis of Aphantasia; Table S15. Others.

Author Contributions: F.J., study design, investigation, methodology, visualization, writing—original draft, writing—review and editing; S.-M.H., investigation, writing—review and editing. Y.L., study design, funding acquisition, investigation, methodology, resources, supervision, writing—original draft, writing—review and editing. All authors have read and agreed to the published version of the manuscript.

Funding: This study was funded by BNU-HKBU United International College Research Grants (R72021207, R202102, R202011), and in part by the Guangdong Provincial Key Laboratory IRADS (2022B1212010006, R0400001-22).

Institutional Review Board Statement: Not applicable.

Informed Consent Statement: Not applicable.

Data Availability Statement: Not applicable.

Conflicts of Interest: The authors declare no conflicts of interest.

References

1. Galton, F. Statistics of mental imagery. *Mind* **1880**, *5*, 301–318. [CrossRef]
2. Zeman, A. Aphantasia and hyperphantasia: Exploring imagery vividness extremes. Trends in Cognitive Sciences. *Trends Cogn. Sci.* **2024**, *28*, 467–480. [CrossRef] [PubMed]
3. Keogh, R.; Pearson, J.; Zeman, A. Aphantasia: The science of visual imagery extremes. *Handb. Clin. Neurol.* **2021**, *178*, 277–296. [PubMed]
4. Zeman, A.; Milton, F.; Della Sala, S.; Dewar, M.; Frayling, T.; Gaddum, J.; Hattersley, A.; Heuerman-Williamson, B.; Jones, K.; MacKisack, M.; et al. Phantasia–The psychological significance of lifelong visual imagery vividness extremes. *Cortex* **2020**, *130*, 426–440. [CrossRef] [PubMed]
5. Zeman, A.; Dewar, M.; Della Sala, S. Lives without imagery—Congenital aphantasia. *Cortex* **2015**, *73*, 378–380. [CrossRef]
6. Dawes, A.J.; Keogh, R.; Andrillon, T.; Pearson, J. A cognitive profile of multi-sensory imagery, memory and dreaming in aphantasia. *Sci. Rep.* **2020**, *10*, 10022. [CrossRef]
7. Pearson, J. The human imagination: The cognitive neuroscience of visual mental imagery. *Nat. Rev. Neurosci.* **2019**, *20*, 624–634. [CrossRef]
8. Dijkstra, N.; Bosch, S.E.; van Gerven, M.A. Shared neural mechanisms of visual perception and imagery. *Trends Cogn. Sci.* **2019**, *23*, 423–434. [CrossRef]
9. Luchini, C.; Veronese, N.; Nottegar, A.; Shin, J.I.; Gentile, G.; Granziol, U.; Soysal, P.; Alexinschi, O.; Smith, L.; Solmi, M. Assessing the quality of studies in meta-research: Review/guidelines on the most important quality assessment tools. *Pharm. Stat.* **2021**, *20*, 185–195. [CrossRef]
10. Shaheen, N.; Shaheen, A.; Ramadan, A.; Hefnawy, M.T.; Ramadan, A.; Ibrahim, I.A.; Hassanein, M.E.; Ashour, M.E.; Flouty, O. Appraising systematic reviews: A comprehensive guide to ensuring validity and reliability. *Front. Res. Metr. Anal.* **2023**, *8*, 1268045. [CrossRef]
11. Blomkvist, A. Aphantasia: In search of a theory. *Mind Lang.* **2022**, *38*, 866–888. [CrossRef]
12. Blomkvist, A.; Marks, D.F. Defining and 'diagnosing' aphantasia: Condition or individual difference? *Cortex* **2023**, *169*, 220–234. [CrossRef] [PubMed]
13. Keogh, R.; Pearson, J. The blind mind: No sensory visual imagery in aphantasia. *Cortex* **2018**, *105*, 53–60. [CrossRef] [PubMed]
14. Fulford, J.; Milton, F.; Salas, D.; Smith, A.; Simler, A.; Winlove, C.; Zeman, A. The neural correlates of visual imagery vividness—An fMRI study and literature review. *Cortex* **2018**, *105*, 26–40. [CrossRef] [PubMed]
15. Jacobs, C.; Schwarzkopf, D.S.; Silvanto, J. Visual working memory performance in aphantasia. *Cortex* **2018**, *105*, 61–73. [CrossRef]
16. Milton, F.; Fulford, J.; Dance, C.; Gaddum, J.; Heuerman-Williamson, B.; Jones, K.; Knight, K.F.; MacKisack, M.; Winlove, C.; Zeman, A. Behavioral and neural signatures of visual imagery vividness extremes: Aphantasia versus hyperphantasia. *Cereb. Cortex Commun.* **2021**, *2*, tgab035. [CrossRef]
17. Bainbridge, W.A.; Pounder, Z.; Eardley, A.F.; Baker, C.I. Quantifying aphantasia through drawing: Those without visual imagery show deficits in object but not spatial memory. *Cortex* **2021**, *135*, 159–172. [CrossRef]
18. Nanay, B. Unconscious mental imagery: Unconscious mental imagery. *Philos. Trans. R. Soc. B Biol. Sci.* **2021**, *376*, 20190689. [CrossRef]
19. Keogh, R.; Pearson, J. Attention driven phantom vision: Measuring the sensory strength of attentional templates and their relation to visual mental imagery and aphantasia: Measuring attentional templates. *Philos. Trans. R. Soc. B Biol. Sci.* **2021**, *376*, 20190688. [CrossRef]
20. Monzel, M.; Mitchell, D.; Macpherson, F.; Pearson, J.; Zeman, A. Aphantasia, dysikonesia, anauralia: Call for a single term for the lack of mental imagery—Commentary on dance et al. (2021) and Hinwar and lambert (2021). *Cortex* **2022**, *150*, 149–152. [CrossRef]
21. Krempel, R.; Monzel, M. Aphantasia and involuntary imagery. *Conscious. Cogn.* **2024**, *120*, 103679. [CrossRef] [PubMed]

22. Marks, D.F. Visual imagery differences in the recall of pictures. *Br. J. Psychol.* **1973**, *64*, 17–24. [CrossRef] [PubMed]
23. Keogh, R.; Pearson, J. Revisiting the blind mind: Still no evidence for sensory visual imagery in individuals with aphantasia. *Neurosci. Res.* **2024**, *201*, 27–30. [CrossRef] [PubMed]
24. Marks, D.F. New directions for mental imagery research. *J. Ment. Imag.* **1995**, *19*, 153–167.
25. Kay, L.; Keogh, R.; Andrillon, T.; Pearson, J. The pupillary light response as a physiological index of aphantasia, sensory and phenomenological imagery strength. *eLife* **2022**, *11*, e72484. [CrossRef]
26. Konigsmark, V.T.; Bergmann, J.; Reeder, R.R. The Ganzflicker experience: High probability of seeing vivid and complex pseudo-hallucinations with imagery but not aphantasia. *Cortex* **2021**, *141*, 522–534. [CrossRef]
27. Reeder, R.R. Ganzflicker reveals the complex relationship between visual mental imagery and pseudo-hallucinatory experiences: A replication and expansion. *Collabra Psychol.* **2022**, *8*, 36318. [CrossRef]
28. Faw, B. Conflicting intuitions may be based on differing abilities: Evidence from mental imaging research. *J. Conscious. Stud.* **2009**, *16*, 45–68.
29. Dance, C.J.; Ipser, A.; Simner, J. The prevalence of aphantasia (imagery weakness) in the general population. *Conscious. Cogn.* **2022**, *97*, 103243. [CrossRef]
30. Monzel, M.; Vetterlein, A.; Reuter, M. No general pathological significance of aphantasia: An evaluation based on criteria for mental disorders. *Scand. J. Psychol.* **2023**, *64*, 314–324. [CrossRef]
31. Takahashi, J.; Saito, G.; Omura, K.; Yasunaga, D.; Sugimura, S.; Sakamoto, S.; Horikawa, T.; Gyoba, J. Diversity of aphantasia revealed by multiple assessments of visual imagery, multisensory imagery, and cognitive style. *Front. Psychol.* **2023**, *14*, 1174873. [CrossRef] [PubMed]
32. Gouveia, L.A. *Imagination Capacity in University Students: Aphantasia and Its Possible Changes in Personal Development*; Seven Editora: São José dos Pinhais, Brazil, 2023. [CrossRef]
33. Beran, M.J.; James, B.T.; French, K.; Haseltine, E.L.; Kleider-Offutt, H.M. Assessing aphantasia prevalence and the relation of self-reported imagery abilities and memory task performance. *Conscious. Cogn.* **2023**, *113*, 103548. [CrossRef] [PubMed]
34. Gulyas, E.; Gombos, F.; Sutori, S.; Lovas, A.; Ziman, G.; Kovacs, I. Visual imagery vividness declines across the lifespan. *Cortex* **2022**, *154*, 365–374. [CrossRef] [PubMed]
35. Campos, A. Gender differences in imagery. *Personal. Individ. Differ.* **2014**, *59*, 107–111. [CrossRef]
36. Aydin, Ç. Gender differences in visual imagery: Object and spatial imagery. *Dokuz Eylül Üniversitesi Sos. Bilim. Enstitüsü Derg.* **2020**, *22*, 1045–1064. [CrossRef]
37. Smyth RS, D.; Acharya, P.; Hunt, N.P. Is visual imagery ability higher for orthodontic students than those in other disciplines? A cross-sectional questionnaire-based study. *J. Orthod.* **2019**, *46*, 205–211. [CrossRef]
38. Fielding, D.; Kahui, V.; Wesselbaum, D. Visual imagination and the performance of undergraduate economics students. *N. Z. Econ. Pap.* **2020**, *54*, 127–137. [CrossRef]
39. Blazhenkova, O.; Pechenkova, E. The two eyes of the blind mind: Object vs. spatial aphantasia? *Russ. J. Cogn. Sci.* **2019**, *6*, 51–65. [CrossRef]
40. Blajenkova, O.; Kozhevnikov, M.; Motes, M.A. Object-spatial imagery: A new self-report imagery questionnaire. *Appl. Cogn. Psychol.* **2006**, *20*, 239–263. [CrossRef]
41. Thorudottir, S.; Sigurdardottir, H.M.; Rice, G.E.; Kerry, S.J.; Robotham, R.J.; Leff, A.P.; Starrfelt, R. The architect who lost the ability to imagine: The cerebral basis of visual imagery. *Brain Sci.* **2020**, *10*, 59. [CrossRef]
42. Knowles, L.; Jones, K.; Zeman, A. Acquired aphantasia in 88 cases: A preliminary report. *J. Neurol. Neurosurg. Psychiatry* **2021**, *92*, A6–A7. [CrossRef]
43. Day, J.; Frayling, T.; Wood, A.; Zeman, A. 130 Does visual imagery vividness have a genetic basis? A genome-wide associa-tion study of 1019 individuals. *J. Neurol. Neurosurg. Psychiatry* **2022**, *93*, A51. [CrossRef]
44. Bumgardner, A.L.; Yuan, K.; Chiu, A.V. I cannot picture it in my mind: Acquired aphantasia after autologous stem cell transplantation for multiple myeloma. *Oxf. Med. Case Rep.* **2021**, *8*, 158–160. [CrossRef] [PubMed]
45. Gaber TA, K.; Eltemamy, M. Post-COVID-19 aphantasia. *Prog. Neurol. Psychiatry* **2021**, *25*, 16–17. [CrossRef]
46. De Vito, S.; Bartolomeo, P. Refusing to imagine? on the possibility of psychogenic aphantasia. A commentary on Zeman et al. (2015). *Cortex* **2016**, *74*, 334–335. [CrossRef]
47. Zeman, A.; Dewar, M.; Della Sala, S. Reflections on aphantasia. *Cortex* **2016**, *74*, 336–337. [CrossRef]
48. Dance, C.J.; Ward, J.; Simner, J. What is the link between mental imagery and sensory sensitivity? insights from aphantasia. *Perception* **2021**, *50*, 757–782. [CrossRef]
49. Takahashi, J.; Gyoba, J. A preliminary single-case study of aphantasia in Japan. *Tohoku Psychol. Folia* **2021**, *79*, 26–32.
50. Bouyer, L.N.; Arnold, D.H. Deep Aphantasia: A visual brain with minimal influence from priors or inhibitory feedback? *Front. Psychol.* **2024**, *15*, 1374349. [CrossRef]
51. Hinwar, R.P.; Lambert, A.J. Anauralia: The silent mind and its association with aphantasia. *Front. Psychol.* **2021**, *12*, 744213. [CrossRef]
52. Wittmann, B.C.; Satirer, Y. Decreased associative processing and memory confidence in aphantasia. *Learn. Mem.* **2022**, *29*, 412–420. [CrossRef] [PubMed]
53. Pounder, Z.; Eardley, A.F.; Loveday, C.; Evans, S. No clear evidence of a difference between individuals who self-report an absence of auditory imagery and typical imagers on auditory imagery tasks. *PLoS ONE* **2024**, *19*, e0300219. [CrossRef] [PubMed]

54. Dawes, A.J.; Keogh, R.; Pearson, J. Multisensory subtypes of aphantasia: Mental imagery as supramodal perception in reverse. *Neurosci. Res.* **2024**, *201*, 50–59. [CrossRef] [PubMed]
55. Ganczarek, J.; Żurawska-Żyła, R.; Rolek, A. "I remember things, but I can't picture them." What can a case of aphantasia tell us about imagery and memory? *Psychiatr. Psychol. Klin.* **2020**, *20*, 134–141. [CrossRef]
56. Keogh, R.; Wicken, M.; Pearson, J. Visual working memory in aphantasia: Retained accuracy and capacity with a different strategy. *Cortex* **2021**, *143*, 237–253. [CrossRef] [PubMed]
57. Knight, K.F.; Milton, F. Memory without Imagery: No Evidence of Visual Working Memory Impairment in People with Aphantasia. In Proceedings of the Annual Meeting of the Cognitive Science Society, Toronto, ON, Canada, 27–30 July 2022; Volume 44.
58. Pounder, Z.; Jacob, J.; Evans, S.; Loveday, C.; Eardley, A.F.; Silvanto, J. Only minimal differences between individuals with congenital aphantasia and those with typical imagery on neuropsychological tasks that involve imagery. *Cortex* **2022**, *148*, 180–192. [CrossRef]
59. Dando, C.J.; Nahouli, Z.; Hart, A.; Pounder, Z. Real-world implications of aphantasia: Episodic recall of eyewitnesses with aphantasia is less complete but no less accurate than typical imagers. *R. Soc. Open Sci.* **2023**, *10*, 231007. [CrossRef]
60. Dawes, A.J.; Keogh, R.; Robuck, S.; Pearson, J. Memories with a blind mind: Remembering the past and imagining the future with aphantasia. *Cognition* **2022**, *227*, 105192. [CrossRef]
61. Monzel, M.; Vetterlein, A.; Reuter, M. Memory deficits in aphantasics are not restricted to autobiographical memory—Perspectives from the dual coding approach. *J. Neuropsychol.* **2022**, *16*, 444–461. [CrossRef]
62. Watkins, N.W. (A)phantasia and severely deficient autobiographical memory: Scientific and personal perspectives. *Cortex* **2018**, *105*, 41–52. [CrossRef]
63. Siena, M.J.; Simons, J. Metacognitive Awareness and the Subjective Experience of Remembering in Aphantasia. *J. Cogn. Neurosci.* **2024**, *36*, 1578–1598. [CrossRef] [PubMed]
64. Kozhevnikov, M.; Kosslyn, S.; Shephard, J. Spatial versus object visualizers: A new characterization of visual cognitive style. *Mem. Cogn.* **2005**, *33*, 710–726. [CrossRef] [PubMed]
65. Crowder, A. Differences in Spatial Visualization Ability and Vividness of Spatial Imagery between People with and without Aphantasia. Doctoral Dissertation, Virginia Commonwealth University, Richmond, VA, USA, 2018.
66. Furman, M.; Fleitas-Rumak, P.; Lopez-Segura, P.; Furman, M.; Tafet, G.; de Erausquin, G.A.; Ortiz, T. Cortical activity involved in perception and imagery of visual stimuli in a subject with aphantasia. an EEG case report. *Neurocase* **2022**, *28*, 344–355. [CrossRef] [PubMed]
67. Palermo, L.; Boccia, M.; Piccardi, L.; Nori, R. Congenital lack and extraordinary ability in object and spatial imagery: An investigation on sub-types of aphantasia and hyperphantasia. *Conscious. Cogn.* **2022**, *103*, 103360. [CrossRef] [PubMed]
68. Keogh, R.; Pearson, J. The perceptual and phenomenal capacity of mental imagery. *Cognition* **2017**, *162*, 124–132. [CrossRef]
69. Shepard, R.N.; Metzler, J. Mental rotation of three-dimensional objects. *Science* **1971**, *171*, 701–703. [CrossRef]
70. Zhao, B.; Della Sala, S.; Zeman, A.; Gherri, E. Spatial transformation in mental rotation tasks in aphantasia. *Psychon. Bull. Rev.* **2022**, *29*, 2096–2107. [CrossRef]
71. Kay, L.; Keogh, R.; Pearson, J. Slower but more accurate mental rotation performance in aphantasia linked to differences in cognitive strategies. *Conscious. Cogn.* **2024**, *121*, 103694. [CrossRef]
72. Eardley, A.F.; Pring, L. Spatial processing, mental imagery, and creativity in individuals with and without sight. *Eur. J. Cogn. Psychol.* **2007**, *19*, 37–58. [CrossRef]
73. Monzel, M.; Keidel, K.; Reuter, M. Imagine, and you will find—Lack of attentional guidance through visual imagery in aphantasics. *Atten. Percept. Psychophys.* **2021**, *83*, 2486–2497. [CrossRef]
74. Monzel, M.; Reuter, M. Where's Wanda? the influence of visual imagery vividness on visual search speed measured by means of hidden object pictures. *Atten. Percept. Psychophys.* **2024**, *86*, 22–27. [CrossRef] [PubMed]
75. Pearson, J.; Clifford, C.W.; Tong, F. The functional impact of mental imagery on conscious perception. *Curr. Biol.* **2008**, *18*, 982–986. [CrossRef] [PubMed]
76. Moriya, J. Visual mental imagery influences attentional guidance in a visual-search task. *Atten. Percept. Psychophys.* **2018**, *80*, 1127–1142. [CrossRef] [PubMed]
77. Speed, L.J.; Eekhof, L.S.; Mak, M. The role of visual imagery in story reading: Evidence from aphantasia. *Conscious. Cogn.* **2024**, *118*, 103645. [CrossRef]
78. Wicken, M.; Keogh, R.; Pearson, J. The critical role of mental imagery in human emotion: Insights from fear-based imagery and aphantasia. *Proc. R. Soc. B-Biol. Sci.* **2021**, *288*, 20210267. [CrossRef] [PubMed]
79. Monzel, M.; Keidel, K.; Reuter, M. Is it really empathy? the potentially confounding role of mental imagery in self-reports of empathy. *J. Res. Personal.* **2023**, *103*, 104354. [CrossRef]
80. Hashim, S.; Pulcini, C.; Jansari, A.; Küssner, M.B.; Omigie, D. The Experience of Music in Aphantasia: Emotion, Reward, and Everyday Functions. *Music Sci.* **2024**, *7*, 20592043231216259. [CrossRef]
81. Keogh, R.; Wicken, M.; Pearson, J. Fewer intrusive memories in aphantasia: Using the trauma film paradigm as a laboratory model of PTSD. *PsyArxiv* **2023**. [CrossRef]
82. Monzel, M.; Dance, C.; Azañón, E.; Simner, J. Aphantasia within the framework of neurodivergence: Some preliminary data and the curse of the confidence gap. *Conscious. Cogn.* **2023**, *115*, 103567. [CrossRef]

83. Cavedon-Taylor, D. Aphantasia and psychological disorder: Current connections, defining the imagery deficit and future directions. *Front. Psychol.* **2022**, *13*, 822989. [CrossRef]
84. Sitek, E.J.; Konieczna, S. Does progressive aphantasia exist? the hypothetical role of aphantasia in the diagnosis of neurodegenerative diseases. *Behav. Brain Sci.* **2022**, *45*, 49–51. [CrossRef] [PubMed]
85. Pearson, J.; Naselaris, T.; Holmes, E.A.; Kosslyn, S.M. Mental imagery: Functional mechanisms and clinical applications. *Trends Cogn. Sci.* **2015**, *19*, 590–602. [CrossRef] [PubMed]
86. Yehuda, R.; Hoge, C.W.; McFarlane, A.C.; Vermetten, E.; Lanius, R.A.; Nievergelt, C.M.; Hobfoll, S.E.; Koenen, K.C.; Neylan, T.C.; Hyman, S.E. Post-traumatic stress disorder. *Nat. Rev. Dis. Primers* **2015**, *1*, 1–22. [CrossRef] [PubMed]
87. Dance, C.J.; Jaquiery, M.; Eagleman, D.M.; Porteous, D.; Zeman, A.; Simner, J. What is the relationship between aphantasia, synaesthesia and autism? *Conscious. Cogn.* **2021**, *89*, 103087. [CrossRef]
88. Crespi, B.; Leach, E.; Dinsdale, N.; Mokkonen, M.; Hurd, P. Imagination in human social cognition 2021, autism, and psychotic-affective conditions. *Cognition* **2016**, *150*, 181–199. [CrossRef]
89. Lord, C.; Elsabbagh, M.; Baird, G.; Veenstra-Vanderweele, J. Autism spectrum disorder. *Lancet* **2018**, *392*, 508–520. [CrossRef]
90. Dance, C.J.; Hole, G.; Simner, J. The role of visual imagery in face recognition and the construction of facial composites. Evidence from Aphantasia. *Cortex* **2023**, *167*, 318–334. [CrossRef]
91. Liu, J.; Bartolomeo, P. Probing the unimaginable: The impact of aphantasia on distinct domains of visual mental imagery and visual perception. *Cortex* **2023**, *166*, 338–347. [CrossRef]
92. Svart, N.; Starrfelt, R. Is it just face blindness? exploring developmental comorbidity in individuals with Self-Reported developmental prosopagnosia. *Brain Sci.* **2022**, *12*, 230. [CrossRef]
93. Monzel, M.; Vetterlein, A.; Hogeterp, S.A.; Reuter, M. No increased prevalence of prosopagnosia in aphantasia: Visual recognition deficits are small and not restricted to faces. *Perception* **2023**, *52*, 03010066231180712. [CrossRef]
94. Spagna, A.; Hajhajate, D.; Liu, J.; Bartolomeo, P. Visual mental imagery engages the left fusiform gyrus, but not the early visual cortex: A meta-analysis of neuroimaging evidence. *Neurosci. Biobehav. Rev.* **2021**, *122*, 201–217. [CrossRef] [PubMed]
95. Dijkstra, N. Uncovering the Role of the Early Visual Cortex in Visual Mental Imagery. *Vision* **2024**, *8*, 29. [CrossRef] [PubMed]
96. Monzel, M.; Leelaarporn, P.; Lutz, T.; Schultz, J.; Brunheim, S.; Reuter, M.; McCormick, C. Hippocampal-occipital connectivity reflects autobiographical memory deficits in aphantasia. *bioRxiv* **2023**. [CrossRef]
97. Dupont, W.; Papaxanthis, C.; Madden-Lombardi, C.; Lebon, F. Explicit and implicit motor simulations are impaired in individuals with aphantasia. *Brain Commun.* **2024**, *6*, fcae072. [CrossRef] [PubMed]
98. Dupont, W.; Papaxanthis, C.; Lebon, F.; Madden-Lombardi, C. Mental simulations and action language are impaired in individuals with aphantasia. *J. Cogn. Neurosci.* **2024**, *36*, 261–271. [CrossRef]
99. Bartolomeo, P. The relationship between visual perception and visual mental imagery: A reappraisal of the neuropsychological evidence. *Cortex* **2002**, *38*, 357–378. [CrossRef]
100. Bartolomeo, P.; Hajhajate, D.; Liu, J.; Spagna, A. Assessing the causal role of early visual areas in visual mental imagery. *Nat. Rev. Neurosci.* **2020**, *21*, 517. [CrossRef]
101. Bartolomeo, P. Visual agnosia and imagery after Lissauer. *Brain* **2021**, *144*, 2557–2559. [CrossRef]
102. Bartolomeo, P.; Bachoud-Lévi, A.C.; De Gelder, B.; Denes, G.; Dalla Barba, G.; Brugières, P.; Degos, J.D. Multiple-domain dissociation between impaired visual perception and preserved mental imagery in a patient with bilateral extrastriate lesions. *Neuropsychologia* **1998**, *36*, 239–249. [CrossRef]
103. Bartolomeo, P. The neural correlates of visual mental imagery: An ongoing debate. *Cortex* **2008**, *44*, 107–108. [CrossRef]
104. Moro, V.; Berlucchi, G.; Lerch, J.; Tomaiuolo, F.; Aglioti, S.M. Selective deficit of mental visual imagery with intact primary visual cortex and visual perception. *Cortex* **2008**, *44*, 109–118. [CrossRef] [PubMed]
105. Riley, S.N.; Davies, J. Vividness as the similarity between generated imagery and an internal model. *Brain Cogn.* **2023**, *169*, 105988. [CrossRef] [PubMed]
106. Arnold, D.H.; Andresen, I.; Anderson, N.; Saurels, B.W. Commonalities between the berger rhythm and spectra differences driven by cross-modal attention and imagination. *Conscious. Cogn.* **2023**, *107*, 103436. [CrossRef] [PubMed]
107. Zeman, A.Z.; Della Sala, S.; Torrens, L.A.; Gountouna, V.E.; McGonigle, D.J.; Logie, R.H. Loss of imagery phenomenology with intact visuo-spatial task performance: A case of 'blind imagination'. *Neuropsychologia* **2010**, *48*, 145–155. [CrossRef] [PubMed]
108. Hajhajate, D.; Kaufmann, B.C.; Liu, J.; Siuda-Krzywicka, K.; Bartolomeo, P. The connectional anatomy of visual mental imagery: Evidence from a patient with left occipito-temporal damage. *Brain Struct. Funct.* **2022**, *227*, 3075–3083. [CrossRef]
109. Spagna, A.; Heidenry, Z.; Miselevich, M.; Lambert, C.; Eisenstadt, B.E.; Tremblay, L.; Liu, Z.; Liu, J.; Bartolomeo, P. Visual mental imagery: Evidence for a heterarchical neural architecture. *Phys. Life Rev.* **2024**, *48*, 113–131. [CrossRef]
110. Liu, J.; Zhan, M.; Hajhajate, D.; Spagna, A.; Dehaene, S.; Cohen, L.; Bartolomeo, P. Ultra-high field fMRI of visual mental imagery in typical imagers and aphantasic individuals. *bioRxiv* **2023**. [CrossRef]
111. Boccia, M.; Sulpizio, V.; Bencivenga, F.; Guariglia, C.; Galati, G. Neural representations underlying mental imagery as unveiled by representation similarity analysis. *Brain Struct. Funct.* **2021**, *226*, 1511–1531. [CrossRef]
112. McCarthy, G.; Puce, A.; Gore, J.C.; Allison, T. Face-specific processing in the human fusiform gyrus. *J. Cogn. Neurosci.* **1997**, *9*, 605–610. [CrossRef]
113. Pearson, J. Reply to: Assessing the causal role of early visual areas in visual mental imagery. *Nat. Rev. Neurosci.* **2020**, *21*, 517–518. [CrossRef]

114. Kosslyn, S.M.; Pascual-Leone, A.; Felician, O.; Camposano, S.; Keenan, J.P.; Thompson, W.L.; Ganis, G.; Sukel, K.E.; Alpert, N.M. The role of area 17 in visual imagery: Convergent evidence from PET and rTMS. *Science* **1999**, *284*, 167–170. [CrossRef] [PubMed]
115. Bergmann, J.; Genç, E.; Kohler, A.; Singer, W.; Pearson, J. Smaller primary visual cortex is associated with stronger, but less precise mental imagery. *Cereb. Cortex* **2016**, *26*, 3838–3850. [CrossRef] [PubMed]
116. Keogh, R.; Bergmann, J.; Pearson, J. Cortical excitability controls the strength of mental imagery. *eLife* **2020**, *9*, e50232. [CrossRef] [PubMed]
117. Meng, M.; Chang, S.; Zhang, X.; Pearson, J. Imageless imagery in aphantasia: Decoding non-sensory imagery in aphantasia. *Res. Sq.* 2023; *preprint*. [CrossRef]
118. Cabbai, G.; Racey, C.; Simner, J.; Dance, C.; Ward, J.; Forster, S. Sensory representations in primary visual cortex are not sufficient for subjective imagery. *bioRxiv* **2024**. [CrossRef]
119. Arcangeli, M. Aphantasia demystified. *Synthese* **2023**, *201*, 31. [CrossRef]
120. Monzel, M.; Handlogten, J.; Reuter, M. No verbal overshadowing in aphantasia: The role of visual imagery for the verbal overshadowing effect. *Cognition* **2024**, *245*, 105732. [CrossRef]
121. Park, T. Why successful performance in imagery tasks does not require the manipulation of mental imagery. *Avant* **2019**, *10*, 1–11. [CrossRef]
122. Toftness, A.R. Clarifying Aphantasia. Doctoral Dissertation, Iowa State University, Ames, IA, USA, 2022.
123. Purkart, R.; Delem, M.; Cavalli, E.; Ranson, V.; Andrey, C.; Versace, R.; Plancher, G. Are There Unconscious Visual Images in Aphantasia? Development of an Implicit Priming Paradigm. Available online: https://ssrn.com/abstract=4755823 (accessed on 10 September 2024).
124. Perrin, D.; Michaelian, K. Memory as mental time travel. In *The Routledge Handbook of Philosophy of Memory*; Routledge: New York, NY, USA, 2017; pp. 228–239.
125. Ungerleider, L.G.; Haxby, J.V. 'What'and 'where'in the human brain. *Curr. Opin. Neurobiol.* **1994**, *4*, 157–165. [CrossRef]
126. Goodale, M.A.; Milner, A.D. Separate visual pathways for perception and action. *Trends Neurosci.* **1992**, *15*, 20–25. [CrossRef]
127. Milner, A.D.; Goodale, M.A. Two visual systems re-viewed. *Neuropsychologia* **2008**, *46*, 774–785. [CrossRef]
128. de Haan, E.H.F.; Jackson, S.R.; Schenk, T. Where are we now with 'What' and 'How'? *Cortex* **2018**, *98*, 1–7. [CrossRef] [PubMed]
129. Grünbaum, T. The two visual systems hypothesis and contrastive underdetermination. *Synthese* **2021**, *198* (Suppl. S17), S4045–S4068. [CrossRef]
130. Bergmann, J.; Ortiz-Tudela, J. Feedback signals in visual cortex during episodic and schematic memory retrieval and their potential implications for aphantasia. *Neurosci. Biobehav. Rev.* **2023**, *152*, 105335. [CrossRef] [PubMed]
131. Suica, Z.; Behrendt, F.; Gäumann, S.; Gerth, U.; Schmidt-Trucksäss, A.; Ettlin, T.; Schuster-Amft, C. Imagery ability assessments: A cross-disciplinary systematic review and quality evaluation of psychometric properties. *BMC Med.* **2022**, *20*, 166. [CrossRef] [PubMed]
132. Schwarzkopf, D.S. What is the true range of mental imagery? *Cortex* **2024**, *170*, 21–25. [CrossRef]
133. Larner, A.J.; Leff, A.P.; Nachev, P.C. Phantasia, aphantasia, and hyperphantasia: Empirical data and conceptual considerations. *Neurosci. Biobehav. Rev.* **2024**, *164*, 105819. [CrossRef]
134. Monzel, M.; Mitchell, D.; Macpherson, F.; Pearson, J.; Zeman, A. Proposal for a consistent definition of aphantasia and hyperphantasia: A response to Lambert and Sibley (2022) and Simner and Dance (2022). *Cortex* **2022**, *152*, 74–76. [CrossRef]
135. Lambert, A.J.; Sibley, C.G. On the importance of consistent terminology for describing sensory imagery and its absence: A response to Monzel et al.(2022). *Cortex* **2022**, *152*, 153–156. [CrossRef]
136. Simner, J.; Dance, C.J. Dysikonesia or aphantasia? understanding the impact and history of names. A reply to monzel et al. (2022). *Cortex* **2022**, *153*, 220–223. [CrossRef]
137. D'Angiulli, A. Vividness, Consciousness and Mental Imagery: A Start on Connecting the Dots. *Brain Sci.* **2020**, *10*, 500. [CrossRef]
138. Küssner, M.B.; Taruffi, L. Modalities and causal routes in music-induced mental imagery. *Trends Cogn. Sci.* **2023**, *27*, 114–115. [CrossRef] [PubMed]
139. Lau, H.; Michel, M.; LeDoux, J.E.; Fleming, S.M. The mnemonic basis of subjective experience. *Nat. Rev. Psychol.* **2022**, *1*, 479–488. [CrossRef]
140. Muraki, E.J.; Speed, L.J.; Pexman, P.M. Insights into embodied cognition and mental imagery from aphantasia. *Nat. Rev. Psychol.* **2023**, *2*, 591–605. [CrossRef]
141. Cohen, M.X. Where does EEG come from and what does it mean? *Trends Neurosci.* **2017**, *40*, 208–218. [CrossRef] [PubMed]
142. Baillet, S. Magnetoencephalography for brain electrophysiology and imaging. *Nat. Neurosci.* **2017**, *20*, 327–339. [CrossRef] [PubMed]
143. Xie, S.; Kaiser, D.; Cichy, R.M. Visual imagery and perception share neural representations in the alpha frequency band. *Curr. Biol.* **2020**, *30*, 2621–2627. [CrossRef]
144. Cichy, R.M.; Oliva, A. A M/EEG-fMRI fusion primer: Resolving human brain responses in space and time. *Neuron* **2020**, *107*, 772–781. [CrossRef]
145. Kriegeskorte, N.; Mur, M.; Bandettini, P.A. Representational similarity analysis-connecting the branches of systems neuroscience. *Front. Syst. Neurosci.* **2008**, *2*, 249. [CrossRef]
146. Haynes, J.D. A primer on pattern-based approaches to fMRI: Principles, pitfalls, and perspectives. *Neuron* **2015**, *87*, 257–270. [CrossRef]

147. Polanía, R.; Nitsche, M.A.; Ruff, C.C. Studying and modifying brain function with non-invasive brain stimulation. *Nat. Neurosci.* **2018**, *21*, 174–187. [CrossRef]
148. Monzel, M.; Karneboge, J.; Reuter, M. The role of dopamine in visual imagery—An experimental pharmacological study. *J. Neurosci. Res.* **2024**, *102*, e25262. [CrossRef] [PubMed]
149. Fried, E.I. Theories and models: What they are, what they are for, and what they are about. *Psychol. Inq.* **2020**, *31*, 336–344. [CrossRef]
150. Gigerenzer, G. Personal reflections on theory and psychology. *Theory Psychol.* **2010**, *20*, 733–743. [CrossRef]
151. Cavedon-Taylor, D. Mental imagery: Pulling the plug on perceptualism. *Philos. Stud.* **2021**, *178*, 3847–3868. [CrossRef]
152. Cavedon-Taylor, D. Untying the knot: Imagination, perception and their neural substrates. *Synthese* **2021**, *199*, 7203–7230. [CrossRef]
153. Cavedon-Taylor, D. Predictive processing and perception: What does imagining have to do with it? *Conscious. Cogn.* **2022**, *106*, 103419. [CrossRef]
154. D'Aloisio-Montilla, N. Imagery and overflow: We see more than we report. *Philos. Psychol.* **2017**, *30*, 545–570. [CrossRef]
155. Duch, W. Imagery agnosia and its phenomenology. *Ann. Psychol.* **2021**, *24*, 307–323. [CrossRef]
156. Fox-Muraton, M. Aphantasia and the language of imagination: A wittgensteinian exploration. *Anal. I Egzystencja* **2021**, *55*, 5–24. [CrossRef]
157. Fox-Muraton, M. A world without imagination? consequences of aphantasia for an existential account of self. *Hist. Eur. Ideas* **2021**, *47*, 414–428. [CrossRef]
158. Lorenzatti, J.J. Aphantasia: A philosophical approach. *Philos. Psychol.* **2023**, 1–29. [CrossRef]
159. MacKisack, M.; Aldworth, S.; Macpherson, F.; Onians, J.; Winlove, C.; Zeman, A. Plural imagination: Diversity in mind and making. *Art J.* **2022**, *81*, 70–87. [CrossRef]
160. Whiteley CM, K. Aphantasia, imagination and dreaming. *Philos. Stud.* **2021**, *178*, 2111–2132. [CrossRef]
161. Open Science Collaboration. Estimating the reproducibility of psychological science. *Science* **2015**, *349*, aac4716. [CrossRef] [PubMed]
162. Tackett, J.L.; Brandes, C.M.; King, K.M.; Markon, K.E. Psychology's replication crisis and clinical psychological science. *Annu. Rev. Clin. Psychol.* **2019**, *15*, 579–604. [CrossRef] [PubMed]
163. Nosek, B.A.; Hardwicke, T.E.; Moshontz, H.; Allard, A.; Corker, K.S.; Dreber, A.; Fidler, F.; Hilgard, J.; Struhl, M.K.; Nuijten, M.B.; et al. Replicability, robustness, and reproducibility in psychological science. *Annu. Rev. Psychol.* **2022**, *73*, 719–748. [CrossRef] [PubMed]
164. Miłkowski, M.; Hensel, W.M.; Hohol, M. Replicability or reproducibility? On the replication crisis in computational neuroscience and sharing only relevant detail. *J. Comput. Neurosci.* **2018**, *45*, 163–172. [CrossRef]
165. Balas, B. Mid-level Characteristics of Drawings Made by Observers with Aphantasia. *J. Percept. Imaging* **2024**, *7*, 1–8. [CrossRef]

vision

MDPI

Review

Shaping the Space: A Role for the Hippocampus in Mental Imagery Formation

Andrea Blomkvist

Centre for the Study of Perceptual Experience, Department of Philosophy, University of Glasgow, Glasgow G12 8QQ, UK; andrea.blomkvist@glasgow.ac.uk

Abstract: Mental imagery is claimed to underlie a host of abilities, such as episodic memory, working memory, and decision-making. A popular view holds that mental imagery relies on the perceptual system and that it can be said to be 'vision in reverse'. Whereas vision exploits the bottom-up neural pathways of the visual system, mental imagery exploits the top-down neural pathways. But the contribution of some other neural areas remains overlooked. In this article, I explore important contributions of the hippocampus, a neural area traditionally associated with episodic memory, to mental imagery formation. I highlight evidence which supports the view that the hippocampus contributes to the spatial model used for mental imagery and argue that we can distinguish different hippocampal circuits which contribute to different kinds of imagery, such as object imagery, scene imagery, and imagery with a temporal aspect. This has significant upshots for mental imagery research, as it opens a new avenue for further research into the role of the hippocampus in a variety of imagery tasks.

Keywords: mental imagery; hippocampus; episodic memory; spatial model; scene construction theory; aphantasia

1. Introduction

Try to remember your graduation. For many of us, remembering this will evoke a mental image. We visualise what the remembered scene looked like, including details of who was there, what they were wearing, where they were standing, and so on. These mental images can aid memory. If asked about details of the scene, I can mentally scan the scene to find the answer. In short, we evoke a mental image of the scene, and this mental image supports our ability to recall details from the event [1]. Similarly, we can evoke a mental image of a future event which we have not yet experienced, though this tends to be less vivid and not as detailed as images of the past [2].

Notably, mental images are not only generated in the visual modality but exist in a host of modalities including audition, taste, smell, and emotion (see Andrade et al. [3] for a subjective measure of different kinds of imagery). 'Mental imagery' is sometimes used as an overarching term for all kinds of imagery and sometimes more narrowly to refer to 'visual imagery' only. In this article, I use it in the more narrow sense, unless otherwise stated. Moreover, though commonly associated with memory and imagination, the generation of an image is not necessary for any of these experiences, as demonstrated by research into aphantasia—an individual difference in the generation of voluntary visual imagery, which is associated with reduced (but not absent) detailedness in autobiographical memory and future imagination [4–6].

In what follows, I am interested in cases where we do generate a visual image when we imagine the future or recall the past. What neural mechanisms contribute to the generation of this image? Research has long supported the involvement of the visual system in visual imagery generation, where it has been suggested that generating a visual image is like 'vision in reverse' [7]. When generating visual imagery, the suggestion is that the visual system is driven in a top-down way by the prefrontal cortex and that content from memory is recruited to then be processed in a reverse hierarchy in the visual system, from high-level areas to low-level areas. Visual imagery and perception are hence thought to share many of the same neural mechanisms [8].

However, mental imagery research has mostly focused on the role of the visual cortex in mental imagery formation, whereas other neural areas which are potentially implicated in the process are relatively underexplored and little is known about their function. One such area is the hippocampus, traditionally thought to support episodic memory. Though it is often claimed that mental imagery takes its content from episodic memory, it has been pointed out that the role of the hippocampus in mental imagery formation is particularly unclear, and some research might even suggest that its involvement is not necessary [7].

In this article, I aim to clarify the role of the hippocampus in mental imagery formation in humans. Though previous work has focused on the role of the hippocampus in imagination, this work has not discussed the contribution to mental imagery formation specifically. The distinction between imagination and mental imagery is an important one which has long been recognised in philosophy, whereby a mental act of imagining is not necessarily accompanied by mental imagery (see Box 1). I aim to provide a novel insight into the role of the hippocampus for mental imagery formation by drawing on work in neuroscience. I argue that we can already discern different hippocampal circuits involved in generating different kinds of imagery, such as object imagery and scene imagery, and that the hippocampus likely supports the building of a spatial model for mental imagery. This has an upshot for the study of mental imagery as it implies that different tasks put different demands on the hippocampus, just as how different tasks put different demands on various visual areas. I argue that this insight is especially important for the study of special populations.

In Section 2, I first give a brief overview of neural areas which are broadly taken to be associated with the generation of mental imagery, before homing in on the hippocampus. I argue that the hippocampus contributes to the formation of both object imagery and scene imagery by providing a spatial model. This is corroborated by neuroscientific evidence showing different activation patterns, as well as data showing deficits in the spatial coherence of mental imagery in a clinical population. I further argue that the hippocampus could play different roles depending on the temporal direction of imagery. In Section 3, I put forward evidence suggesting that the hippocampus is not involved in mental imagery formation, and I suggest that this conclusion does not follow as other plausible interpretations of the data are available. In Section 4, I discuss the consequences for mental imagery research and suggest that researchers ought to be sensitive to what tasks are set, especially when studying special populations, as hippocampal circuits supporting different kinds of imagery could potentially dissociate. Finally, in Section 5, I conclude that we ought to further investigate the role of the hippocampus in imagery formation, as there are many aspects left to explore.

Box 1. Propositional imagination and sensory imagination.

Philosophers make a distinction between propositional imagination and sensory imagination [9]. Propositional imagination involves imagining that something is the case (e.g., imagining that *Paris is not the capital of France*) whereas sensory imagination involves deploying one's sensorimotor systems to evoke a mental image (e.g., imagining *walking down the street of Paris and hearing the sound of French being spoken*). Mental imagery is taken by many to be a necessary component of sensory imagination, but not of propositional imagination.

2. The Mechanisms of Mental Imagery Generation

The generation of mental imagery is a complex process associated with multiple neural areas. To better contextualise the contributions of the hippocampus, I briefly discuss the contributions of other areas, including prefrontal areas and visual areas (for a detailed discussion, see [7,10]). I also limit my discussion here to the formation of voluntary visual imagery, as this is what most research focuses on.

2.1. Visual Cortex and Prefrontal Areas

The investigation of the neural processes of visual imagery has a long history. Most research has focused on the relationship between visual perception and visual imagery, asking to what extent visual areas support the creation of visual imagery. It has been hypothesised that the creation of visual imagery is like 'vision in reverse' [7], a reverse hierarchy where prefrontal areas drive activation in visual cortices, from higher to lower levels (for a recent discussion, see Dijkstra [11]). Particular attention has been paid to the potential activation of V1, which has been hypothesised to be necessary for the generation of visual imagery ([10] but see also [12,13]), and wherefrom researchers have been able to decode the content of mental imagery [14]. This suggests that there are common patterns of activation for perception and visual imagery, which become increasingly similar at higher cortical levels but which can be detected as early as V1. The prefrontal cortex (PFC) is also highlighted as an area which contributes to mental imagery formation. As this area is active in a wide range of imagery tasks, it is likely that it plays a more executive role in driving the activation of subsequent areas [15–17]. It is also possible that it selects schema on which imagery is scaffolded [18,19].

2.2. Anatomy of the Hippocampus

Before discussing the contributions of the hippocampus to mental imagery formation, it is useful to know about the functional neuroanatomy of the hippocamps. The hippocampus is part of the default mode network (DMN), which is involved in both remembering and imagining scenes and events [20,21]. This network comprises the hippocampal formation (hippocampus, enthorinal cortex, subiculum, presubiculum, parasubiculum and the parahippocampal gyrus [22]. The hippocampus itself can be divided into three parts—head, body, and tail. The body of the hippocampus is divided into cytoarchitectonically defined subfields comprising the dentate gyrus (DG), CA3, CA2, and CA1. These form a closed circuit where information is inputted via the enthorinal cortex, which in turn drives DG, CA3, CA2, and CA1 and the subiculum. The subiculum and CA1 then project back to the EC and the loop is closed. Interestingly, recent research has also demonstrated direct connections between the hippocampus and various cortical areas using in vivo quantitative fibre tracking in humans (for similar work on primates, see [23]). This work has demonstrated preferential connectivity along the anterior-posterior hippocampal axis—for example, between the tail of the hippocampus and occipital areas such as V1 and V2. Areas in the occipital cortex also display higher endpoint density in the posterior medial hippocampus and, to a lesser degree, in the anterior medial hippocampus. These anatomical links could be crucial for mental imagery formation, as discussed later.

In terms of function, the hippocampus has long been an area which we know to be important for episodic memory encoding and retrieval, but research has also shown that it is involved in future episodic thought [20,21], counterfactual thinking [24], navigation [25], as well as perception [26–28]. Research on non-human animals has also contributed greatly to our understanding of the function of the hippocampus, potentially supporting its playing a role in mental imagery generation. For example, research has shown that the dorsal hippocampus in rats plays a role in filling in missing information, potentially as a visual image, of visual cues (for an overview, see [29]). Techniques using a machine interface have also supported the volitional access of cognitive maps to navigate a virtual reality environment in rats, a strategy which is akin to imagining a possible route [30]. However, as pointed out [31], as there are anatomical differences between the human and non-human hippocampus, we need to be careful when generalising about function from one species to another. For the remainder of this article, I limit my discussion to the role hippocampus plays in imagery formation in humans, especially as I want to focus on voluntary imagery, something which is difficult to assess in other animals.

This new research in both human and non-human animals puts pressure on the old idea that the hippocampus is the 'seat of memory' and that its sole function is to support episodic memory. Instead, another suggestion is that the hippocampus is involved in the more general processing of scenes, whether these are perceived, imagined, or remembered [26,32]. Scene construction theory suggests that the hippocampus is responsible for constructing the spatial representation necessary for creating a spatially coherent scene. Supporting research shows that hippocampal lesion leads to an inability to construct spatially coherent scenes [33], and fMRI studies support the involvement of the hippocampus especially in tasks which require a 3D spatial layout [29,30,34–36].

But how does this complex system contribute to mental imagery formation? So far, the emphasis in research has been on how the hippocampus, as part of the DMN, contributes to episodic recall or future imagination by supplying content of past experiences. However, neither of these functions are necessarily accompanied by mental imagery, as we can propositionally imagine the future or remember the past (see Box 1), and many studies do not assess whether participants engage in mental imagery, or whether they merely think about a past or future event without having an accompanied experience of imagery. Luckily, a handful of studies have assessed whether participants engage in mental imagery formation during these tasks by using subjective measures such as vividness ratings. I highlight these studies as informative regarding the role of the hippocampus in mental imagery formation. I argue that, in line with scene construction theory, data suggest that the hippocampus plays the role of building a spatial representation in mental imagery during scene imagery and object imagery. Scene imagery involves visualising a 3D space populated with objects, which one could potentially 'step into', such as a jungle or a living room. Object imagery involves visualising one or multiple objects, but where these are located on a plain background rather than in a wider scene. Further, the hippocampus also contributes to the temporal direction represented in mental imagery [37]. Interestingly, as we will see, the role of the hippocampus in different kinds of imagery formation is nuanced.

2.3. *Hippocampal Contributions to Object Imagery vs. Scene Imagery*

An fMRI study by Dalton et al. [34] investigated the contribution of the hippocampus to scene imagery vs. object imagery. This experiment involved scanning participants whilst they were asked to visualise the following: an object, a 2D grid, arrays of objects placed on a 2D grid, a 3D grid, and objects placed on a 3D grid (referred to as 'scene imagery'). All cases were instructed to be static images, where perspective was not shifted. For all tasks, the participants focused on a white screen with their eyes open and were explicitly told not to recall from memory. For imagining objects, the participants were presented with three separate object descriptions and asked to imagine these objects on a plain background one at a time. For a 2D grid, they were simply asked to imagine a 2D grid covering two-thirds of the white screen. For the array of objects, they were asked to imagine the 2D grid and then 'place' three objects on it at separate locations. For the 3D grid, they were asked to imagine a 3D grid stretching out over two-thirds of the white screen. And finally, for scene imagery, they were first asked to imagine the 3D grid and then 'place' three separate objects in it at different locations. Crucially, the participants rated their vividness of each imagined item on a scale between 1 and 5 (1 being 'not at all', and 5 being 'extremely vivid'), and only cases which were rated as 4 or above were included in the analysis. As Dalton et al. were interested in testing the contributions of the hippocampus in imagining scenes, they were primarily interested in contrasting the cases of imagining an array of objects on a 2D grid with scene imagery. However, for my present purposes, further contrast cases are also of interest, as they are telling in how the hippocampus could potentially contribute in different ways to imagery formation.

The fMRI results showed that the hippocampus, as well as other brain areas, were differentially activated depending on the task. That is, depending on what kind of imagery a participant formed, different subregions of the hippocampus were active. First, contrasting the imagery of objects with scenes revealed bilateral activation in the perirhinal cortex (PRC), entorhinal cortex (ENT), and right anterior lateral hippocampus. Contrasting scene imagery with the imagery of objects instead showed activation in the bilaterial pre/parasubiculum, parahippocampal cortex (PHC), retrosplenial cortex (RSC), and posterior cingulate cortex (PCC). Second, contrasting imagining an array of objects with constructing a scene engaged the bilateral ENT, PRC, early visual cortices (EVC), and left posterior hippocampus. This is consistent with previous studies which have also shown greater activation in the occipital cortex during object imagery [36]. Contrasting constructing a scene with imagining an array of objects instead showed activation in the anterior medial hippocampus, PHC, RSC, and PCC.

The activation of these different areas is in line with previous findings. These studies have shown that portions of the pre/parasubibulum in the anterior medial hippocampus are involved in scene construction [38], and Dalton et al. suggest that these areas support the construction of scenes couched in a naturalistic 3D spatial framework. This is particularly plausible as this area, due to its anatomical location [39], could have access to holistic representations of the environment and could be a 'hub' for scene-based recognition [38]. Evidence from connectivity studies [40] showing that the hippocampus displays greater connectivity to visual areas than previously thought further bolsters this possibility. Based on these results, Dalton et al. concluded that the hippocampus (as well as other areas) is preferentially activated depending on the task and that there are multiple processing circuits within the hippocampus which engage depending on the task demand.

A further study which investigated the temporal dynamics of scene imagery and object imagery using MEG similarly supports the claim that different circuits are involved [17]. The participants were asked to either imagine a 3D space that they could step into (such as a jungle) (there were no instructions regarding whether this should be a static scene or a temporally unfolding scene, so it is possible that the participants could have imagined it as temporally unfolding), imagine a single object on a white background (such as a cushion), or perform a baseline task (counting backwards). They were explicitly told not to recall from memory. The results showed that the left anterior hippocampus showed a significant change from baseline in terms of a theta power decrease in both imagery tasks. These results corroborate Dalton et al.'s (2018) findings that the anterior hippocampus is involved in scene imagery [34,41]. Interestingly, there was also an observed theta increase between vmPFC and the hippocampus, which was larger when imagining scenes than when imagining objects. Barry et al. suggest that this could be due to vmPFC being involved in extracting regularities across scenes to form schemas, which can then serve as 'templates' for novel scenes [18,19]. Hence, here too, we see that the hippocampus (in conjunction with vmPFC) is involved in both scene imagery and object imagery, but these two tasks are supported by different circuits (for a broader discussion on scene and object processing within the hippocampus, see [42]).

Finally, a telling study suggests that lesions to the hippocampus also have consequences for the mental imagery formation of scenes, where a phenomenon known as 'boundary extension' is not experienced by patients with bilateral hippocampal lesion [33]. Boundary extension occurs when we recall parts of a scene which were beyond our actual viewpoint at the time [43], and whether it occurs is now thought to depend on the properties of the stimuli [44]. Unlike the controls, the patients did not experience this effect in a number of paradigms which required them to either draw or report on a previously viewed photograph. When asked what could be beyond the scene that they had previously viewed in the photograph, they were able to recount objects that could be present but unable to say how these would spatially relate to the scene they had seen. These results strongly suggest that the hippocampus is involved in building a coherent spatial representation of a scene and that lesion to the hippocampus impairs this ability.

2.4. Hippocampal Contributions to Mental Imagery with a Temporal Aspect

Mental imagery can also unfold over time and hence contain a temporal aspect. That is to say, some mental imagery can be more like a mental video, where subjects do not just experience one snapshot but rather the unfolding of an event. Some studies have investigated the role of the hippocampus in mental imagery of past and future events which contain this temporal aspect. For example, Addis et al. [37] investigated participants who were tasked with constructing a past event or a future event and contrasted a construction phase of the event with an elaboration phase. A conjunction analysis comparing the imagery of a past and future event to a control task revealed that a number of regions were commonly recruited, including the left hippocampus, right inferior parietal lobule, left superior occipital gyrus/cuneus, and right middle occipital gyrus. However, comparing the construction phase to the elaboration phase did not reveal any significant differences in hippocampal activations but did elicit differences in other brain areas. Hence, the hippocampus might contribute to a similar extent to both the construction phase and the elaboration phase. It is also possible here that different sub-regions of the hippocampus contribute differently, similar to the findings previously discussed, but sub-regions of the hippocampus were not assessed separately in this study. Notably, another temporal aspect that mattered to the areas activated was temporal direction, that is, whether the event was represented as being in the past or the future. For both temporal directions, the left hippocampus was equally activated. However, comparing the imagery of a future event with the imagery of a past event, the right hippocampus was significantly more activated during the creation of future events. This suggest that the temporal direction represented in imagery also elicits different responses for the hippocampus and that it could serve a role in generating non-imagistic content in mental imagery, such as tagging an event as representing the past or future (for a similar idea regarding the format of mental imagery as being 'cartographic, see [45]).

2.5. Summary

What does this tell us about mental imagery formation? As different hippocampal circuits are involved depending on task demand, this suggest that different subregions of the hippocampus (and surrounding areas) make different contributions to various kinds of imagery. Interestingly, this mimics findings in vision science where different areas of the visual cortex are involved in the formation of different kinds of imagery, such as areas associated with object recognition being active for object imagery [10]. However, we have not known that similar patterns due to task demand can be traced in the hippocampus. Scene construction theory could explain these findings by appealing to a more complex spatial layout being required for constructing a 3D scene, compared to constructing an object in a 2D space. Potentially, different circuits support the construction of 3D space vs. 2D space, and hence, we see one circuit being recruited when participants are asked to construct scene imagery and another circuit being recruited when participants are asked to construct object imagery. Another possibility is that the different circuits reflect the complexity of the task, where scene imagery is generally more complex than object imagery as it often involves more features and more complex relations between objects. Moreover, the temporal direction also elicits different responses from the hippocampus, suggesting that the hippocampus could play a role when it comes to representing imagery with different temporal content. As such, it might be more apt to say that the hippocampus provides a spatiotemporal model for imagery formation, rather than only a spatial model.

To sum up, even though few studies have been conducted which specifically investigate the contributions of the hippocampus to mental imagery formation, we can already detect a pattern whereby different subregions of the hippocampus and related extra-hippocampal areas support different imagery tasks. Now, though these studies demonstrate a role for different parts of the hippocampus in the generation of different imagery representations, speculating about the function of these different anatomical parts would be premature, as the evidence so far is fairly scarce, making it difficult to form specific testable hypotheses. Future research should extend these findings—for example, by looking at what demands are placed on the hippocampus when engaging in even more complex imagery formation, such as imagery formation in different modalities (e.g., visual, auditory, emotion) or imagery of future plausible/implausible events [46]. Another possible function of the hippocampus in mental imagery generation could be sampling from a generative model during both perception and imagery [47], where more such sampling would be required for generating a scene compared to a single object. This proposal appears compatible with the data presented in this article, as it posits a neurocomputational model explaining how the hippocampus could achieve the higher-level function of providing a spatiotemporal model. Note that my claims in this article are not about how the function is accomplished on a neurocomputational level. Future research should also further investigate this hypothesis.

3. Evidence Against Hippocampal Involvement in Mental Imagery Generation

Before moving on, I will address two findings which purportedly go against my suggestion in this article. First, Maguire et al. [48] conducted a study which was subsequently reported as failing to find a role for the hippocampus in imagery formation (Pearson, 2019). Maguire et al. report on two amnesic patients (one with acquired amnesia and one with developmental amnesia) and their ability to imagine fictitious events and future experiences. Interestingly, both these patients were able to construct these scenarios to the same levels of detailedness and spatial coherence as controls. This is surprising if the hippocampus is crucial for imagery generation, and especially if it is crucial for generating 3D spatial scenes. Hence, it might be tempting to say that this study shows that the hippocampus is not crucial for scene imagery generation. However, there is a possible explanation for these patients' intact ability, as 50% of the patients' hippocampi remained intact and showed activation on MRI. It is hence possible that they could perform these tasks with support from the residual hippocampi tissue, as suggested by Maguire et al. in the original study. Note that vividness was not explicitly rated in this study; participants instead evaluated *saliency* from 'could not see anything at all' to 'extremely salient', which uses very similar terminology to a standard vividness questionnaire.

Second, another study which could cast doubt on the role of the hippocampus in mental imagery formation [7] tested the ability of amnesic patients with bilateral hippocampal damage regarding spatial transformation tasks [49]. These included mental rotation, imagery for spatial location, mental paper folding, and the Mental Imagery Questionnaire. They found no differences in performance between patients and controls and concluded that the hippocampus is not necessary for forming the mental imagery required for solving these tasks. This seems surprising, as solving these tasks is thought to depend on creating mental imagery with complex spatial relations (e.g., in the mental rotation task). However, other research shows that these tasks can be solved without appealing to mental imagery, as alternative strategies are available. This interpretation is supported by a recent study of participants with aphantasia (defined here as having reduced or no imagery ability) [50], which found that these participants gave more accurate but slower responses to mental rotation tasks. Interestingly, accurate but slower response time is also what Kim et al. reported when testing amnesiac patients. What sheds further light on the issue is that Kay et al. additionally queried participants on what strategy they used to solve the task. This revealed that aphantasics had a preference for analytic, rather than rotation-based, strategies—they were most likely not solving the task by using mental imagery [51].

Hence, mental rotation tasks can be solved without the use of mental imagery, and it is possible that patients in Kim et al.'s study used alternative strategies to solve the task. For the remaining tasks tested by Kim et al., similar possible strategies for solving the tasks were not investigated or ruled out, and it remains possible that these too could be solved by alternative strategies as it is not clear that mental imagery is necessary for any one of them (even if many people may use mental imagery as a strategy). In fact, similar results have been found in other working memory tasks [51]. To conclude, there is no clear evidence against the suggestions that the hippocampus contributes a spatiotemporal model to mental imagery.

4. Upshot for Mental Imagery Research

The involvement of the hippocampus in mental imagery formation is consequential for how we should design tasks and questionnaires to investigate mental imagery. Mental imagery questionnaires are frequently used in neuroscience and psychology to assess mental imagery ability, such as in aphantasia research (for a recent discussion, see [52]) or across a lifespan [53]. Of particular note are two widely used questionnaires: the VVIQ [54,55] and the Plymouth Sensory Imagery Questionnaire (PSI-Q) [3]. These questionnaires comprise questions which elicit scene imagery and object imagery, and some of these questions ask participants to explicitly draw on memory. For example, half of the questions (8/16) of VVIQ explicitly focus on eliciting imagery based on past memory, including a relative or friend that one knows well, and the front of a shop which one frequently visits. The other questions are more generic (a sunset, and a country scene) and do not ask subjects to draw on past memory (though subjects could of course still do so). In contrast, PSI-Q contains 35 questions, of which 5 focus on visual imagery, and 2 of these could be thought to cue memories ("imagine a friend you know well" and "imagine the appearance of the front door of your house"), whereas the others are more generic (e.g., a cat climbing a tree, a bonfire). No questions in non-visual modalities explicitly call on subjects to use their memory. It is also interesting to note that some questions ask participants to imagine a scene (e.g., a country scene from the VVIQ, or a cat climbing a tree from the PSI-Q), whereas others ask them to imagine single objects (e.g., a bonfire from the PSI-Q, or the door of the shopfront from the VVIQ).

Based on the previous discussion, we should expect differential contributions of the hippocampus to these heterogeneous tasks, as some involve explicitly drawing on memory, some involve object imagery, some involve scene imagery, and some also involve different temporal aspects. Though these kinds of questionnaires might give us a good overview of someone's ability to generate mental imagery by testing heterogeneous tasks, they could also potentially obscure differences. For example, if assessing clinical populations (especially populations with hippocampal lesions), we need to pay close attention to what kind of imagery we are asking a patient to form, as it could be the case that a deficit only encompasses one kind of imagery (e.g., scene imagery) even if another kind of imagery (e.g., object imagery) could be formed.

This is not to say that mental imagery should only be assessed by subjective questionnaires, as this method alone cannot tell us about the neural mechanisms of mental imagery formation, and there is also the added concern of subjects being required to introspect their mental state, which could be unreliable [56]. Philosophers have further pointed out that central concepts used in mental imagery research—notably 'vividness'—are poorly understood and not appropriately defined in questionnaires [57]. Tackling these questions is central for mental imagery research to move forward, and interdisciplinary research projects would be an ideal way of making headway here, as skills from philosophy, neuroscience, and psychology are required.

Moreover, paying closer attention to the kind of imagery we ask participants in experiments to form could also reveal new findings in non-clinical populations, such as people with aphantasia. We already have reason to believe that people we currently classify as aphantasic actually make up a heterogeneous group [52], as there are varying reports from people who cannot form imagery in any modality, people who cannot form visual imagery but can form non-visual imagery, and people who can form involuntary but not voluntary visual imagery [58]. Different underlying neural mechanisms could give rise to these different symptoms [59]. Moreover, recent research also points to different hippocampal activity in the aphantasic population compared to controls when retrieving autobiographical memories [6]. As this indicates hippocampal involvement, it warrants asking more nuanced questions as to what aspects of content in imagery people with aphantasia could potentially form. As aphantasia is most commonly defined as reduced, rather than absent, voluntary visual imagery, studies include participants reporting 'vague and dim' imagery in the aphantasic category. However, which aspects of imagery this subgroup (sometimes referred to as 'hypophantasic' [60]) can form has not been investigated. A possible aspect which should be investigated is whether there are differences when it comes to scene imagery, object imagery, and imagery based on the temporal direction between this group and controls. More nuanced thinking about the role of the hippocampus in mental imagery formation should lead us to hypothesise that there could be hitherto unnoticed differences in responses to imagery tasks as a result of the contributions of different hippocampal circuits.

5. Conclusions

I have argued based on neuroscientific evidence that it is likely that the hippocampus plays a role in mental imagery formation by building the spatiotemporal model. Though we still do not know exactly which function different sub-regions perform, evidence suggests that the hippocampus does instantiate different functions depending on task demand, where different circuits are deployed depending on what kind of imagery is formed. This has significant upshots for mental imagery research, which up until now has mostly focused on the role of the visual system in imagery formation. We should further investigate the role that the hippocampus plays, and researchers ought to be sensitive to the hippocampus' differential activation patterns when designing tasks and questionnaires. This is of particular importance when studying special populations, as this could potentially reveal more about the involvement of hippocampal subregions in forming mental imagery.

Funding: This research was funded by the British Academy, grant number PFSS23\230115, and a Lord Kelvin/Adam Smith (LKAS) Fellowship from University of Glasgow.

Institutional Review Board Statement: Not applicable.

Informed Consent Statement: Not applicable.

Data Availability Statement: No new data were created or analysed in this study. Data sharing is not applicable to this article.

Acknowledgments: I am grateful to Derek Brown, Will Hornett, Jack Lyons, and Fiona Macpherson for a very helpful discussion of my manuscript. I am also grateful to four anonymous reviewers for their comments on the manuscript.

Conflicts of Interest: The author declares no conflicts of interest. The funders had no role in the design of the study; in the collection, analyses, or interpretation of the data; in the writing of the manuscript; or in the decision to publish the results.

References

1. Madore, K.P.; Jing, H.G.; Schacter, D.L. Episodic Specificity Induction and Scene Construction: Evidence for an Event Construction Account. *Conscious. Cogn.* **2019**, *68*, 1–11. [CrossRef]
2. Addis, D.R.; Pan, L.; Vu, M.A.; Laiser, N.; Schacter, D.L. Constructive Episodic Simulation of the Future and the Past: Distinct Subsystems of a Core Brain Network Mediate Imagining and Remembering. *Neuropsychologia* **2009**, *47*, 2222–2238. [CrossRef]
3. Andrade, J.; May, J.; Deeprose, C.; Baugh, S.J.; Ganis, G. Assessing Vividness of Mental Imagery: The Plymouth Sensory Imagery Questionnaire. *Br. J. Psychol.* **2014**, *105*, 547–563. [CrossRef] [PubMed]
4. Milton, F.; Fulford, J.; Dance, C.; Gaddum, J.; Heuerman-Williamson, B.; Jones, K.; Knight, K.F.; MacKisack, M.; Winlove, C.; Zeman, A. Behavioral and Neural Signatures of Visual Imagery Vividness Extremes: Aphantasia versus Hyperphantasia. *Cereb. Cortex Commun.* **2021**, *2*, tgab035. [CrossRef] [PubMed]
5. Dawes, A.J.; Keogh, R.; Robuck, S.; Pearson, J. Memories with a Blind Mind: Remembering the Past and Imagining the Future with Aphantasia. *Cognition* **2022**, *227*, 105192. [CrossRef] [PubMed]
6. Monzel, M.; Leelaarporn, P.; Lutz, T.; Schultz, J.; Brunheim, S.; Reuter, M.; Mccormick, C. Hippocampal-Occipital Connectivity Reflects Autobiographical Memory Deficits in Aphantasia. *eLife* **2024**, *13*, RP94916. [CrossRef] [PubMed]
7. Pearson, J. The Human Imagination: The Cognitive Neuroscience of Visual Mental Imagery. *Nat. Rev. Neurosci.* **2019**, *20*, 624–634. [CrossRef] [PubMed]
8. Dijkstra, N.; Bosch, S.E.; van Gerven, M.A.J. Shared Neural Mechanisms of Visual Perception and Imagery. *Trends Cogn. Sci.* **2019**, *23*, 423–434. [CrossRef]
9. Currie, G.; Ravenscroft, I. *Recreative Minds: Imagination in Philosophy and Psychology*; Oxford University Press: Oxford, UK, 2002; Volume 20.
10. Kosslyn, S.M. *Image and Brain: The Resolution of the Imagery Debate*; MIT Press: Cambridge, MA, USA, 1994.
11. Dijkstra, N. Uncovering the Role of the Early Visual Cortex in Visual Mental Imagery. *Vision* **2024**, *8*, 29. [CrossRef] [PubMed]
12. Spagna, A.; Hajhajate, D.; Liu, J.; Bartolomeo, P. Visual Mental Imagery Engages the Left Fusiform Gyrus, but Not the Early Visual Cortex: A Meta-Analysis of Neuroimaging Evidence. *Neurosci. Biobehav. Rev.* **2021**, *122*, 201–217. [CrossRef] [PubMed]
13. Bartolomeo, P.; Hajhajate, D.; Liu, J.; Spagna, A. Assessing the Causal Role of Early Visual Areas in Visual Mental Imagery. *Nat. Rev. Neurosci.* **2020**, *21*, 517. [CrossRef] [PubMed]
14. Johnson, M.R.; Johnson, M.K. Decoding Individual Natural Scene Representations during Perception and Imagery. *Front. Hum. Neurosci.* **2014**, *8*, 59. [CrossRef]
15. Ishai, A.; Ungerleider, L.G.; Haxby, J. V Distributed Neural Systems for the Generation of Visual Images Parts of the Visual System Mediate Imagery for Different Types of Visual Information. Patients with Prosopagnosia Due to Occipitotemporal Damage Were Unable to Imagine. *Neuron* **2000**, *28*, 979–990. [CrossRef]
16. McCormick, C.; Barry, D.N.; Jafarian, A.; Barnes, G.R.; Maguire, E.A. VmPFC Drives Hippocampal Processing during Autobiographical Memory Recall Regardless of Remoteness. *Cereb. Cortex* **2020**, *30*, 5972–5987. [CrossRef]

17. Barry, D.N.; Barnes, G.R.; Clark, I.A.; Maguire, E.A. The Neural Dynamics of Novel Scene Imagery. *J. Neurosci.* **2019**, *39*, 4375–4386. [CrossRef]
18. Gilboa, A.; Marlatte, H. Neurobiology of Schemas and Schema-Mediated Memory. *Trends Cogn. Sci.* **2017**, *21*, 618–631. [CrossRef] [PubMed]
19. van Kesteren, M.T.R.; Beul, S.F.; Takashima, A.; Henson, R.N.; Ruiter, D.J.; Fernández, G. Differential Roles for Medial Prefrontal and Medial Temporal Cortices in Schema-Dependent Encoding: From Congruent to Incongruent. *Neuropsychologia* **2013**, *51*, 2352–2359. [CrossRef] [PubMed]
20. Schacter, D.L.; Addis, D.R. The Cognitive Neuroscience of Constructive Memory: Remembering the Past and Imagining the Future. *Philos. Trans. R. Soc. B Biol. Sci.* **2007**, *362*, 773–786. [CrossRef] [PubMed]
21. Schacter, D.L.; Addis, D.R. Memory and Imagination: Perspectives on Constructive Episodic Simulation. In *The Cambridge Handbook of the Imagination*; Cambridge University Press: Cambridge, UK, 2020; pp. 111–131. [CrossRef]
22. Insausti, R.; Marcos, M.P.; Mohedano-Moriano, A. The Nonhuman Primate Hippocampus: Neuroanatomy and Pattern of Cortical Activity. In *The Hippocampus from Cells to Systems Structure, Connectivity, and Functional Contributions to Memory and Flexible Cognition*; Hannula, D.E., Duff, M.C., Eds.; Springer: Cham, Switzerland, 2017.
23. Insausti, R.; Muñoz, M. Cortical Projections of the Non-Entorhinal Hippocampal Formation in the Cynomolgus Monkey (*Macaca fascicularis*). *Eur. J. Neurosci.* **2001**, *14*, 435–451. [CrossRef]
24. De Brigard, F.; Addis, D.R.; Ford, J.H.; Schacter, D.L.; Giovanello, K.S. Remembering What Could Have Happened: Neural Correlates of Episodic Counterfactual Thinking. *Neuropsychologia* **2013**, *51*, 2401–2414. [CrossRef] [PubMed]
25. Epstein, R.A.; Patai, E.Z.; Julian, J.B.; Spiers, H.J. The Cognitive Map in Humans: Spatial Navigation and Beyond. *Nat. Neurosci.* **2017**, *20*, 1504–1513. [CrossRef] [PubMed]
26. Zeidman, P.; Maguire, E.A. Anterior Hippocampus: The Anatomy of Perception, Imagination and Episodic Memory. *Nat. Rev. Neurosci.* **2016**, *17*, 173–182. [CrossRef] [PubMed]
27. Treder, M.S.; Charest, I.; Michelmann, S.; Mart In-Buro, C.; Ed Eric Roux, F.; Carceller-Benito, F.; Ugalde-Canitrot, A.; Rollings, D.T.; Sawlani, V.; Chelvarajah, R.; et al. The Hippocampus as the Switchboard between Perception and Memory. *Proc. Natl. Acad. Sci. USA* **2021**, *118*, e2114171118. [CrossRef] [PubMed]
28. Aitken, F.; Kok, P. Hippocampal Representations Switch from Errors to Predictions during Acquisition of Predictive Associations. *Nat. Commun.* **2022**, *13*, 3294. [CrossRef] [PubMed]
29. Blaisdell, A.P. Mental Imagery in Animals: Learning, Memory, and Decision-Making in the Face of Missing Information. *Learn. Behav.* **2019**, *47*, 193–216. [CrossRef]
30. Lai, C.; Tanaka, S.; Harris, T.D.; Lee, A.K. Volitional Activation of Remote Place Representations with a Hippocampal Brain-Machine Interface. *Science* **2023**, *382*, 566–573. [CrossRef] [PubMed]
31. Fan, C.L.; Sokolowski, H.M.; Rosenbaum, R.S.; Levine, B. What about "Space" Is Important for Episodic Memory? *Wiley Interdiscip. Rev. Cogn. Sci.* **2023**, *14*, e1645. [CrossRef]
32. Mullally, S.L.; Maguire, E.A. Memory, Imagination, and Predicting the Future: A Common Brain Mechanism? *Neuroscientist* **2014**, *20*, 220–234. [CrossRef]
33. Mullally, S.L.; Intraub, H.; Maguire, E.A. Attenuated Boundary Extension Produces a Paradoxical Memory Advantage in Amnesic Patients. *Curr. Biol.* **2012**, *22*, 261–268. [CrossRef]
34. Dalton, M.A.; Zeidman, P.; McCormick, C.; Maguire, E.A. Differentiable Processing of Objects, Associations, and Scenes within the Hippocampus. *J. Neurosci.* **2018**, *38*, 8146–8159. [CrossRef]
35. McCormick, C.; Dalton, M.A.; Zeidman, P.; Maguire, E.A. Characterising the Hippocampal Response to Perception, Construction and Complexity. *Cortex* **2021**, *137*, 1–17. [CrossRef]
36. Hassabis, D.; Kumaran, D.; Maguire, E.A. Using Imagination to Understand the Neural Basis of Episodic Memory. *J. Neurosci.* **2007**, *27*, 14365–14374. [CrossRef] [PubMed]
37. Addis, D.R.; Wong, A.T.; Schacter, D.L. Remembering the Past and Imagining the Future: Common and Distinct Neural Substrates during Event Construction and Elaboration. *Neuropsychologia* **2007**, *45*, 1363–1377. [CrossRef] [PubMed]
38. Dalton, M.A.; Maguire, E.A. The Pre/Parasubiculum: A Hippocampal Hub for Scene-Based Cognition? *Curr. Opin. Behav. Sci.* **2017**, *17*, 34–40. [CrossRef]
39. Kravitz, D.J.; Saleem, K.S.; Baker, C.I.; Mishkin, M. A New Neural Framework for Visuospatial Processing. *Nat. Rev. Neurosci.* **2011**, *12*, 217–230. [CrossRef] [PubMed]
40. Dalton, M.A.; D'souza, A.; Lv, J.; Calamante, F. New Insights into Anatomical Connectivity along the Anterior–Posterior Axis of the Human Hippocampus Using in Vivo Quantitative Fibre Tracking. *eLife* **2022**, *11*, e76143. [CrossRef] [PubMed]
41. Clark, I.A.; Kim, M.; Maguire, E.A. Verbal Paired Associates and the Hippocampus: The Role of Scenes. *J. Cogn. Neurosci.* **2018**, *30*, 1821–1845. [CrossRef] [PubMed]
42. Grande, X.; Sauvage, M.; Becke, A.; Düzel, E. Transversal Functional Connectivity and Scene-1 Specific Processing in the Human Entorhinal-2 Hippocampal Circuitry 3 4. *eLife* **2022**, *11*, e76479. [CrossRef]

43. Intraub, H.; Bender, R.S.; Mangels, J.A. Looking at Pictures But Remembering Scenes. *J. Exp. Psychol. Learn. Mem. Cogn.* **1992**, *18*, 180–191. [CrossRef] [PubMed]
44. Bainbridge, W.A.; Baker, C.I. Boundaries Extend and Contract in Scene Memory Depending on Image Properties. *Curr. Biol.* **2020**, *30*, 537–543.e3. [CrossRef] [PubMed]
45. Clarke, S. Mapping the Visual Icon. *Philos. Q.* **2022**, *72*, 552–577. [CrossRef]
46. Miceli, K.; Morales-Torres, R.; Khoudary, A.; Faul, L.; Parikh, N.; De Brigard, F. Perceived Plausibility Modulates Hippocampal Activity in Episodic Counterfactual Thinking. *Hippocampus* **2024**, *34*, 2–6. [CrossRef] [PubMed]
47. Gershman, S.J. The Generative Adversarial Brain. *Front. Artif. Intell.* **2019**, *2*, 18. [CrossRef] [PubMed]
48. Maguire, E.A.; Vargha-Khadem, F.; Hassabis, D. Imagining Fictitious and Future Experiences: Evidence from Developmental Amnesia. *Neuropsychologia* **2010**, *48*, 3187–3192. [CrossRef] [PubMed]
49. Kim, S.; Borst, G.; Thompson, W.L.; Hopkins, R.O.; Kosslyn, S.M.; Squire, L.R. Sparing of Spatial Mental Imagery in Patients with Hippocampal Lesions. *Learn. Mem.* **2013**, *20*, 657–663. [CrossRef] [PubMed]
50. Kay, L.; Keogh, R.; Pearson, J. Slower but More Accurate Mental Rotation Performance in Aphantasia Linked to Differences in Cognitive Strategies. *Conscious. Cogn.* **2024**, *121*, 103694. [CrossRef]
51. Keogh, R.; Wicken, M.; Pearson, J. Visual Working Memory in Aphantasia: Retained Accuracy and Capacity with a Different Strategy. *Cortex* **2021**, *143*, 237–253. [CrossRef]
52. Blomkvist, A.; Marks, D.F. Defining and 'Diagnosing' Aphantasia: Condition or Individual Difference? *Cortex* **2023**, *169*, 220–234. [CrossRef] [PubMed]
53. Gulyás, E.; Gombos, F.; Sütöri, S.; Lovas, A.; Ziman, G.; Kovács, I. Visual Imagery Vividness Declines across the Lifespan. *Cortex* **2022**, *154*, 365–374. [CrossRef]
54. Marks, D.F. New Directions for Mental Imagery Research. *J. Ment. Imag.* **1995**, *19*, 153–167.
55. Marks, D.F. Visual Imagery Differences in the Recall of Pictures. *Br. J. Psychol.* **1973**, *64*, 17–24. [CrossRef] [PubMed]
56. Schwitzgebel, E. How Well Do We Know Our Own Conscious Experience? The Case of Visual Imagery. *J. Conscious. Stud.* **2002**, *9*, 35–53.
57. Kind, A. Imaginative Vividness. *J. Am. Philos. Assoc.* **2017**, *3*, 32–50. [CrossRef]
58. Dawes, A.J.; Keogh, R.; Andrillon, T.; Pearson, J. A Cognitive Profile of Multi-Sensory Imagery, Memory and Dreaming in Aphantasia. *Sci. Rep.* **2020**, *10*, 10022. [CrossRef] [PubMed]
59. Blomkvist, A. Aphantasia: In Search of a Theory. *Mind Lang.* **2022**, *38*, 866–888. [CrossRef]
60. Königsmark, V.T.; Bergmann, J.; Reeder, R.R. The Ganzflicker Experience: High Probability of Seeing Vivid and Complex Pseudo-Hallucinations with Imagery but Not Aphantasia. *Cortex* **2021**, *141*, 522–534. [CrossRef] [PubMed]

vision

MDPI

Article

Brain Functional Connectivity During First- and Third-Person Visual Imagery

Ekaterina Pechenkova [1,*], Mary Rachinskaya [1], Varvara Vasilenko [1,2], Olesya Blazhenkova [3] and Elena Mershina [1,4]

[1] Laboratory for Cognitive Research, School of Psychology, Faculty of Social Sciences, HSE University, 101000 Moscow, Russia; vvasil5664@gmail.com (V.V.)
[2] Institute of Social Sciences, Russian Presidential Academy of National Economy and Public Administration, 119571 Moscow, Russia
[3] Faculty of Arts and Social Sciences, Sabanci University, 34956 Istanbul, Turkey; olesya@sabanciuniv.edu
[4] University Clinic, Lomonosov Moscow State University, 119991 Moscow, Russia
* Correspondence: epechenkova@hse.ru

Abstract: The ability to adopt different perspectives, or vantage points, is fundamental to human cognition, affecting reasoning, memory, and imagery. While the first-person perspective allows individuals to experience a scene through their own eyes, the third-person perspective involves an external viewpoint, which is thought to demand greater cognitive effort and different neural processing. Despite the frequent use of perspective switching across various contexts, including modern media and in therapeutic settings, the neural mechanisms differentiating these two perspectives in visual imagery remain largely underexplored. In an exploratory fMRI study, we compared both activation and task-based functional connectivity underlying first-person and third-person perspective taking in the same 26 participants performing two spatial egocentric imagery tasks, namely imaginary tennis and house navigation. No significant differences in activation emerged between the first-person and third-person conditions. The network-based statistics analysis revealed a small subnetwork of the early visual and posterior temporal areas that manifested stronger functional connectivity during the first-person perspective, suggesting a closer sensory recruitment loop, or, in different terms, a loop between long-term memory and the "visual buffer" circuits. The absence of a strong neural distinction between the first-person and third-person perspectives suggests that third-person imagery may not fully decenter individuals from the scene, as is often assumed.

Keywords: mental imagery; visual imagery; perspective; vantage point; first-person perspective; third-person perspective; fMRI; functional connectivity

1. Introduction

1.1. Perspective Taking, and First-Person Versus Third-Person Perspectives

Vantage point, or perspective, refers to the position from which an individual views or mentally represents a scene. People can adopt different vantage points when encoding, retrieving and using spatial information [1]. This ability may be fundamental to the development of human thinking, as it supports mental object manipulation, understanding others' beliefs, and reflexive reasoning. Adult individuals can mentally shift their imagined perspectives depending on the demands of a task. The ability to shift away from an egocentric viewpoint and to consider multiple aspects of a situation emerges in the course of a child's cognitive development and was described by Jean Piaget as decentration [2].

Besides reasoning, decentration and perspective taking are intensively involved in memory and imagination. The ability to shift between personal and external perspectives influences how events are recalled and communicated.

Autobiographical memory research distinguishes between first-person memories (often called field-perspective in the memory literature), where events are recalled as if re-experiencing them through one's own eyes, and third-person (observer) memories, where individuals recall events as if watching themselves from an external vantage point [3]. First-person recall is often associated with a stronger sensory and emotional reliving of the event. Third-person recall tends to be used in self-distancing strategies, reducing emotional intensity and promoting reflection [4].

During imagery, individuals also can adopt either a first-person (representing space from their own viewpoint) or third-person (external observer) perspective when visualizing scenes or events [5]. Notably, besides decentering/detaching from the first-person vantage point, third-person perspective imagery allows visualizing one's own body in a mental image constructed from an external viewpoint.

Imagining a scene from a third-person perspective should require more mental effort compared to a first-person perspective in both of the following respects: first, additional cognitive load may be needed to reconstruct the scene from an unfamiliar viewpoint [6]; second, integrating one's own body position with a scene viewed externally may also demand extra resources, as suggested by an electrophysiological study showing more pronounced neural correlates of cognitive effort in the frontal lobes in third-person compared to first-person imagery [7].

Third-person imagery may introduce more spatial distortions compared to first-person imagery. Since we naturally experience the world from our own eyes, projecting ourselves into an external viewpoint can lead to miscalculations in spatial relationships [8]. Additionally, subjective body ownership, a sense of embodiment, and reactions to threat are stronger during the first-person perspective than the third-person perspective [9]. Visuomotor information is more effectively conveyed through a first-person perspective versus a third-person perspective. Also, latency to imitate an action shown in video-clips is shorter during the first-person perspective compared to the third-person perspective [10,11].

Notably, the first-person vs. third-person distinction is different from the concept of allocentric and egocentric coordinate systems. The egocentric coordinate system defines locations relative to the observer's body (e.g., left-right, near-far). This system is closely linked to action-oriented processing and self-movement [12]. The allocentric coordinate system defines locations relative to external objects or landmarks, independent of the observer's current position [13]. While allocentric and egocentric spatial processing are governed by distinct cognitive and neurological processes [14,15], they also interact dynamically in many spatial tasks. Perspective-taking tasks, as well as most real-world navigation, way-finding, and spatial orientation, require egocentric processing since they involve maintaining and updating self-referenced spatial relationships. Additionally, even though common perspective-taking tasks [16] include a map, which provides a global (allocentric) layout, study participants must simulate an embodied presence within the map, making the task predominantly egocentric. If the task required participants to view the spatial layout from an external, bird's-eye perspective and indicate directions between objects independently of their own position, it would be more allocentric. However, when imagining oneself from the outside, a person is more likely to simply change one egocentric coordinate system to another, moving its origin to the location of another subject (imagined embodied human observer), than to switch to an allocentric system [17].

Modern visual media (e.g., video games, movies, virtual reality) frequently present experiences from a third-person perspective or allow users to switch between perspectives (e.g., navigators in vehicles). This might make the practice of remembering and imagining oneself from the outside (a third-person perspective) more common. From a neuroscience perspective, it raises an interesting question about what these practices might change in the human brain. Given the importance of decentration for the development of human thought, surprisingly little is known about how different mental vantage points are represented in the brain during internally generated experiences. Revealing the underlying mechanisms of imagery perspectives may not only advance theoretical knowledge but also promote education (e.g., [18]), immersive media [19], and psychotherapy [20].

1.2. Brain Underpinnings of First-Person and Third-Person Perspectives

We are aware of only one published study directly comparing the neural correlates of the same imagery tasks performed from the first- and third-person perspectives, which is a PET study by Ruby and Decety [21]. When the participants were performing imagery tasks from the third-person perspective, more pronounced activation was recorded in the medial prefrontal cortex, the precuneus, and the lateral parietal cortex, which are now known to belong to the default-mode network (DMN). In an early unpublished fMRI study [17], participants were asked to count objects seen from their own perspective and from the perspective of an avatar; "externalizing" the perspective was accompanied by activation in the right lateral parietal cortex, and counting from one's own perspective was accompanied by activation in the medial parietal and prefrontal cortex. Meanwhile, Tomasino et al. [22] found no difference in activation when asking participants to imagine situations described in first- and third-person sentences.

An ERP study by Arzy et al. [23] showed that, in healthy people, embodied imaginary self-location is associated with the greater activity in the extrastriate body area (EBA), and disembodied imaginary self-location is associated with greater activity in the temporoparietal junction (TPJ). Another study [24] demonstrated that transcranial magnetic stimulation over the TPJ may disrupt the imagery process when healthy participants try to imagine an out-of-body experience. However, an fMRI study of a clinical case involving a participant who could voluntarily have an out-of-body experience showed that, compared to imagining herself from the outside, the out-of-body experience was accompanied by increased activation in the lingual gyrus (visual cortex) and decreased activation in the orbitofrontal cortex, but it did not involve any greater activation of the TPJ [25].

Several more recent neurocognitive studies were conducted in the field of autobiographical memory. A resting-state fMRI study has shown that seeing one's own body during episodic encoding modulates the subsequent resting-state functional connectivity between the right hippocampal formation and the neocortex [26]. In one direct task-based fMRI comparison, the observer perspective was associated with greater activity in the right precuneus and in the right TPJ [27], while in another study, observer memories were accompanied by significantly decreased activity of the insulae and the left somatomotor cortex compared to field memories [28]. Based on these and some other relevant findings, a neurocognitive model suggested by St. Jacques [29] incorporates two brain subsystems supporting the visual perspective during episodic memory retrieval: the bodily self subsystem, including the somatosensory cortex, the insular cortex, and the right TPJ; and the viewpoint-specific mental imagery subsystem, including the angular gyrus and the precuneus.

To summarize, the available neuroimaging data indicate a wide range of potential brain mechanisms for externalizing the perspective during imagery. However, none of the available studies have directly compared the task-based functional connectivity (FC) associated with first-person and third-person perspectives during mental visual imagery in the brain. To obtain more straightforward evidence related to the difference in the brain underpinnings of first-person and third-person self-imagery, we conducted an experiment in which the same participants were asked to perform visual and motor imagery tasks taking two perspectives—the first-person and the third-person—during fMRI scanning. In the two imagery conditions, we compared whole-brain activation and task-based FC.

We used two fMRI paradigms developed by the Coma Science Group and widely used as imagery tasks in healthy volunteers and in clinical populations: imaginary tennis and imaginary navigation around one's house [30]. Comparing the neural correlates of the vantage points across the two tasks allows for disentangling the invariant effects of perspective from the task-specific mechanisms. The fact that some previous neuroimaging results indicate the involvement of components of the DMN, which is sensitive to task difficulty, together with other findings indicating that one perspective may take greater mental effort than another [7], we asked our participants to report at the end of the experiment which of the two perspectives they considered more difficult.

While we expected possible greater signs of embodied cognition in the first-person perspective (stronger activity or FC in the bodily self subsystem, according to St. Jacques [29]), given the heterogeneous results of the previous few fMRI studies, we applied an exploratory approach and looked for any potential task-based activation differences in FC.

2. Materials and Methods

2.1. Participants

Twenty-eight healthy volunteers from the academic community of Moscow took part in the study. Inclusion criteria were age 18–45 y.o. and Russian being the native language; exclusion criteria were self-reported history of neurological or psychiatric disease, hearing problems, and contraindications to MRI. All participants passed the MRI safety screening and provided written informed consent before the experiment. Handedness was assessed with the Edinburgh Handedness Inventory by Oldfield [31].

Data from two participants were excluded from the analysis (one male, one female; one for not following the instructions and opening their eyes during the scanning session, and another for not hearing the task-switching commands due to technical reasons), resulting in 26 participants in the final sample (18 females, 8 males; mean age 24 $\pm$ 6 y.o.). Their mean laterality quotient (LQ10) was 75 $\pm$ 21, and all participants were right-handed with the exception of two who were ambidextrous (0 < LQ10 < 40).

2.2. Procedure

Participants were not trained to perform the tasks, but they received detailed instructions before the scanning. Two imagery tasks, tennis and house navigation, were explained. For the tennis task, the experimenter asked the participant to imagine being on a tennis court, serving, and returning subsequent shots in a game with an imaginary opponent. If the participants had no personal experience of playing tennis, they were asked to imagine badminton, which is a more popular outdoor activity in Russia. The participants were asked to start imagining the game at the 'Play tennis' command and to keep the game unraveling in their mind until the 'Stop' command. For the house navigation task, the experimenter asked the participant to imagine wandering across a familiar house or apartment with several rooms (preferably their own or family home) and visualizing the furnishings of these rooms and the objects in them along the way. The participants were asked to start navigating at the 'Walk around the house' command and to keep up this mental activity until the 'Stop' command.

Then, the difference between the first-person and the third-person perspectives was explained. Participants were instructed to look at the imaginary activities either from the vantage point of their own eyes or to look at themselves as seen by an outside observer, as if they were watching a movie about themselves. A conditional visual illustration of these scenes is presented in Figure 1.

Figure 1. A visual illustration of the scenes implied in the imagery tasks: (**a**) the house navigation task with third-person perspective; (**b**) the house navigation task with first-person perspective; (**c**) the tennis task with third-person perspective; (**d**) the tennis task with first-person perspective. The pictures were generated using FLUX Tools by Black Forest Labs (https://blackforestlabs.ai/ (accessed on 1 March 2025)).

In the scanner, all tasks were performed with eyes closed, and the experimenter only asked the participants to open their eyes in between the functional sessions to prevent them from falling asleep.

For each participant, four sessions of functional imaging were performed, one per combination of the imagery task (tennis or house navigation) and viewpoint (first or third perspective). The order of the four conditions (sessions) was counterbalanced across participants using a Latin square design. Participants were informed of the upcoming task and the required mental viewpoint before the start of each session. Within each session, the task was organized in 16 s blocks alternating with 16 s resting periods (rest, or the baseline condition, lasted from the 'Stop' command to the next imagery command). The order of blocks in sessions is illustrated in Figure 2. There were 8 task blocks and 9 resting blocks in each session, which is a typical session length for the Coma Science Group paradigms (e.g., see [32] suggesting comparable results for 5 to 10 blocks for the 1.5T and 5 blocks for 3T scanners).

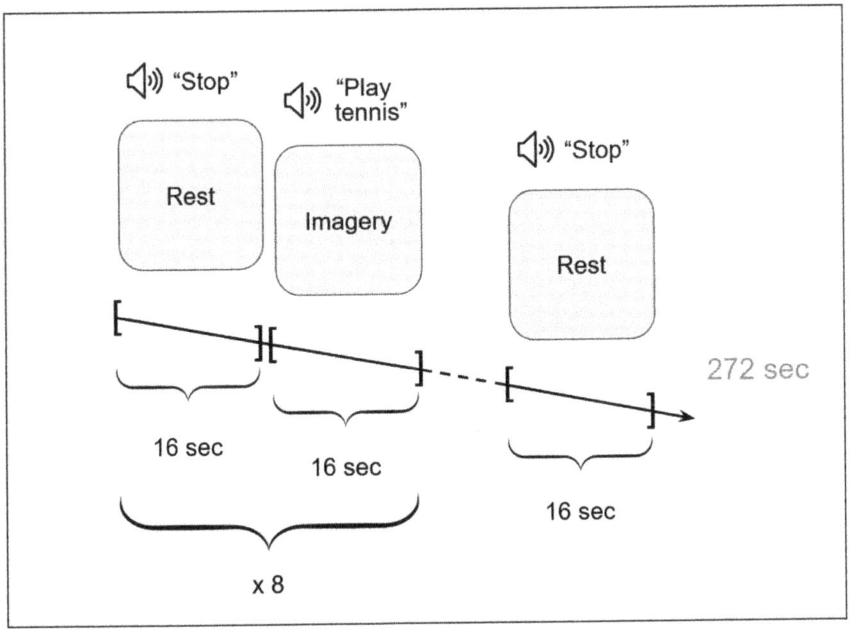

Figure 2. The order of task and rest blocks in the 'Play tennis' session.

At the end of the experiment, the participants were asked which of the two perspectives was more difficult for them to imagine.

2.3. Functional MRI Data Acquisition and Analysis

2.3.1. Functional MRI Data Acquisition

MRI data were acquired with a 1.5T Magnetom Avanto scanner (Siemens, Erlangen, Germany) located at the Medical Research and Educational Centre, Lomonosov Moscow State University in Moscow, Russia, which is equipped with a standard Head Matrix 12-channel coil. During the scanning sessions, instructions to participants were provided via SensaVue (InVivo Corporation, Gainesville, FL, USA).

In each of the four functional sessions, 140 T2*-weighted functional images were acquired, and 4 initial volumes in each session were discarded. Sessions lasted for 4 min 46 sec each. The following parameters were specified for the gradient-echo echo-planar imaging (GRE EPI) pulse sequence: TR/TE/FA = 2000 ms/50 ms/83°, FoV = 256 mm × 256 mm × 96 mm, matrix size = 64 × 64 × 22, slice gap 5%, resulting in an isotropic voxel size of 4 mm. Each volume covered the whole brain with slices oriented parallel to the AC/PC line. For some participants, the total coverage of the brain volume with the specified FoV was not possible, so the inferior portions of the cerebellum were not imaged. Field maps with the same slice prescriptions as the functional images were acquired at the midpoint of the experiment with a standard Siemens double-echo GRE field mapping sequence (TR/TE1/TE2 = 460 ms/4.76 ms/9.52 ms). T1-weighted MPR structural images with 1 mm isotropic resolution were also acquired for each participant.

2.3.2. Functional MRI Data Analysis: Activation

Data were preprocessed with SPM 12 v. 7771 (Wellcome Institute of Cognitive Neurology, www.fil.ion.ucl.ac.uk/spm (accessed on 1 March 2025)) and the following steps: slice timing correction; realignment and fieldmap-based unwarping; spatial co-registration of the structural and functional images; segmentation of the average structural volume into six tissue volumes; normalization into Montreal Neurological Institute (MNI) space; and spatial smoothing of the functional images with a Gaussian kernel of 8 mm full width at half-maximum (FWHM). Six residual head motion parameters (three for translation and three for rotation) were extracted during the realignment step.

To reveal the task-based activation, for each participant, the MR signal in each voxel was modeled using the general linear model with canonical hemodynamic response function. The data were analyzed as a block design with one experimental condition (imagery task) per session, while the baseline condition (rest) was not explicitly modeled to avoid model redundancy. Six parameters describing head motion throughout the experimental session were included into the model as nuisance regressors. *T*-test contrasts for the BOLD signal change evoked by imagery were obtained for each session (i.e., task × perspective combination). Four contrast images per individual (one per condition) entered a flexible-factorial second-level SPM model with a subject factor introduced to control for repeated measures. First, the group activation maps were constructed for all imagery conditions vs. baseline (rest) to reveal brain areas involved in closed-eye visual imagery in general. Next, the between-task contrasts were assessed across both perspectives to reveal the neural correlates specific to each imagery task (tennis and house navigation). Finally, the between-perspective contrasts were estimated across both tasks and within each task individually. The results were assessed with a topological FDR correction for multiple comparisons, with a cluster-wise threshold of $p_{FDR} < 0.05$ (q = 0.05) based on a voxel-wise threshold of $p = 0.001$ uncorrected. Peak activation coordinates were labeled with the Harvard–Oxford maximal likelihood cortical and subcortical atlases in the version implemented in Conn [33].

2.3.3. Functional MRI Data Analysis: Functional Connectivity

CONN Functional Connectivity Toolbox v. 22a was used for the task-based functional connectivity statistical analysis. First, the data initially preprocessed with SPM12 for the activation analysis entered several standard denoising Conn procedures [34]. In order to account for the residual motion-induced artifacts and physiological noise, head motion artifact detection was performed with the Artifact Detection Toolbox (ART; [35]) at medium-level thresholds; outliers were identified as images demonstrating scan-to-scan head motion of more than 0.9 mm or global mean intensity change of more than 5 SDs. Next, an anatomical component-based noise correction technique (aCompCor; [36]) was applied; noise ROI were defined within the individual CSF and white matter masks for each participant. Head motion parameters, outliers, CSF/white matter signal, main BOLD-signal effects of the imagery task and rest blocks, and a linear detrending term were included into the denoising model and regressed out. Finally, a standard temporal high-pass filter with a cutoff of 0.008 Hz was applied in order to restrict the analysis to the frequency band which is characteristic of the task-based fMRI BOLD signal.

Two exploratory approaches to the task-based functional connectivity analysis were taken: voxel-to-voxel and entire atlas ROI-to-ROI. The Harvard–Oxford atlas in the version implemented in the Conn toolbox (105 labels) was used for the ROI definition and results labeling.

For a voxel-to-voxel analysis, the intrinsic connectivity contrast (ICC) values were computed for each voxel in the whole brain. ICC was obtained as a mean absolute value of the correlations of the time series for a given voxel with all other voxels included in the analysis [37]. As the ICC coefficients were further used for within-subject comparisons, the across-brain normalization was not applied.

For both exploratory analyses, three contrasts were estimated: the first-person vs. third-person perspectives were compared across both tasks and within each task individually. The mean-centered percent of the valid scans per subject was included into the model as a between-subject covariate in order to account for possible task-related individual differences in residual participant head motion.

Differences in ICC were assessed with both parametric and non-parametric statistics for cluster-based inference with an FDR correction for multiple comparisons [38,39]. Differences in ROI-to-ROI connectivity were assessed with a non-parametric network-based statistics (NBS) analysis [40]. NBS provides an analog to the cluster-based inference for the ROI-to-ROI analyses, i.e., a set of statistical techniques for testing hypotheses about the clusters of connections (networks) rather than individual connections. The multiple comparison correction in NBS is implemented similarly to topological (cluster-wise) family-wise error/false discovery rate (FWE/FDR) corrections for the voxel-based maps. For the current study, we used NBS by intensity (mass) statistics with a connection-level threshold of $p < 0.001$ uncorrected and a cluster-level threshold of $p < 0.05$, FDR-corrected at the network level, and looked for any network (set of clustered connections) that showed a significant change in connectivity between the first-person and third-person perspective in both imagery tasks or in each task separately.

3. Results

3.1. Subjective Reports

Among 26 participants, 10 reported that the first-person condition was less difficult, 7 experienced it as more difficult, 3 people reported that the first-person perspective required less mental effort for the house navigation task and more effort for the tennis task, and 6 people reported both imagery vantage points as being of equal difficulty.

3.2. Brain Activation

Figure 3 presents the overall activation map of areas manifesting significantly greater or significantly less BOLD signal in both imagery tasks and in both perspectives compared with the baseline (rest), with relevant cluster-wise statistics shown in Table 1. Regions that were more activated in imagery vs. baseline included bilateral frontal and parietal areas typically assigned to the frontoparietal control network (middle and superior frontal gyri, supplementary motor area, superior parietal lobule), the bilateral occipitoparietal areas, and the basal ganglia. Regions more activated in baseline vs. imagery included the bilateral operculum, auditory cortex, occipital cortex, angular gyrus, postcentral gyrus, posterior cingulate, and medial prefrontal cortex.

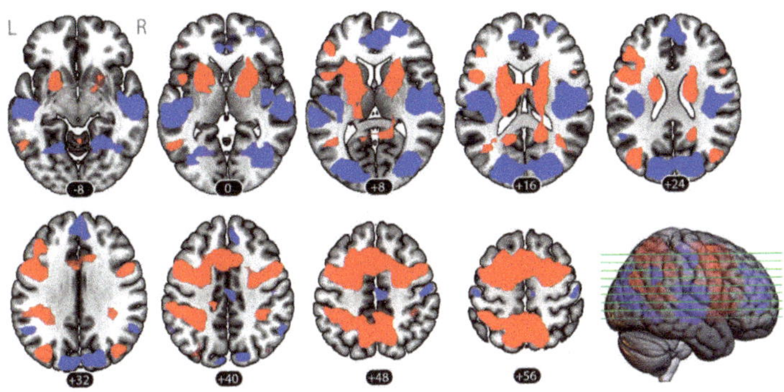

Figure 3. Overall neural correlates of closed-eye imagery tasks compared to baseline (rest): group map, $n = 26$, voxel-level threshold $p < 0.001$, cluster-wise threshold $p_{FDR} < 0.05$. Activation (imagery > baseline) is shown in red; deactivation (baseline > imagery) appears in blue. The statistical map is overlaid onto an MNI template. All data are presented in the MNI space.

Table 1. Clusters of activation characteristic to both imagery tasks and perspectives in all participants.

Cluster	Volume, Voxels (mm^3)	p_{FDR}	MNI Coordinates (Center of Mass)			Region Labels [1]
			x	y	z	
Imagery > Rest (Baseline)						
(1) Frontal Bilateral	1961 (125,504)	<0.001	−7	2	32	Middle Frontal Gyrus (MidFG), LR Precentral Gyrus (PreCG),LR Superior Frontal Gyrus (SFG), LR Supplementary Motor Cortex (SMA), LR Thalamus, LR Putamen, LR
(2) Parietal Bilateral	665 (42,560)	<0.001	−14	−52	52	Precuneous Cortex Superior Parietal Lobule (SPL), LR Lateral Occipital Cortex, superior division (sLOC), LR Postcentral Gyrus (PostCG), L Supramarginal Gyrus, posterior division (pSMG), L
(3) Occipito-temporal Left	127 (8128)	<0.001	−42	−68	19	Lateral Occipital Cortex, superior division (sLOC), L Middle Temporal Gyrus, temporooccipital part (toMTG), L
(4) Occipital Right	59 (3776)	0.002	40	−69	25	Lateral Occipital Cortex, superior division (sLOC), R
(5) Parietal Right	33 (2112)	0.020	36	−39	39	Supramarginal Gyrus, posterior division (pSMG), R Superior Parietal Lobule (SPL), R
Rest (Baseline) > Imagery						
(1) Right Operculum	658 (28,032)	<0.001	48	−17	8	Central Opercular Cortex (CO), R Parietal Operculum Cortex (PO), R Insular Cortex (IC), R Middle Temporal Gyrus, posterior division (pMTG), R Superior Temporal Gyrus, posterior division (pSTG), R Heschl's Gyrus (HG), R Planum Temporale (PT), R
(2) Left Operculum	354 (42,112)	<0.001	−49	−23	7	Central Opercular Cortex (CO), L Parietal Operculum Cortex (PO), L Insular Cortex (IC), L Middle Temporal Gyrus, posterior division (pMTG), L Superior Temporal Gyrus, posterior division (pSTG), L Heschl's Gyrus (HG), L Planum Temporale (PT), L Planum OPerculum (PO), L

Table 1. *Cont.*

Cluster	Volume, Voxels (mm^3)	p_{FDR}	MNI Coordinates (Center of Mass)			Region Labels [1]
			x	y	z	
(3) Occipital Bilateral	438 (22,656)	<0.001	5	−79	14	Lateral Occipital Cortex, superior division (sLOC), LR Lateral Occipital Cortex, inferior division (iLOC), LR Cuneal Cortex, LR Occipital Pole (OP), LR Lingual Gyrus (LG), LR
(4) Medial Frontal	281 (17,984)	<0.001	7	48	16	Paracingulate Gyrus (PaCiG), LR Cingulate Gyrus, anterior division (AC) Frontal Pole (FP), R Superior Frontal Gyrus (SFG), LR
(5) Left Angular Gyrus	25 (1600)	0.028	−53	−54	31	Angular Gyrus (AG), L
(6) Right Angular Gyrus	24 (1536)	0.028	55	−54	35	Angular Gyrus (AG), R
(7) Posterior Cingulate	36 (2304)	0.014	4	−19	46	Cingulate Gyrus, posterior division (PC) Precentral Gyrus (PreCG), R
(8) Right Postcentral	32 (2048)	0.017	49	−20	52	Postcentral Gyrus (PostCG), R
(9) Left Postcentral	24 (1536)	0.028	−44	−22	61	Postcentral Gyrus (PostCG), L

[1] For each cluster, only the anatomical labels covering more than 5% of cluster volume are presented; voxel size is 4 mm isotropic.

The between-task contrasts are shown in Figure 4 and Table 2. As demonstrated by the group activation map, the imaginary tennis task elicited greater activation compared to the imaginary house navigation task in the supplementary motor area (SMA), the right supramarginal gyrus, and the left superior parietal cortex. In turn, the imaginary house navigation task resulted in greater engagement of the bilateral hippocampal gyrus, the occipitotemporal cortex, and the frontal eye field regions, compared to the imaginary tennis task.

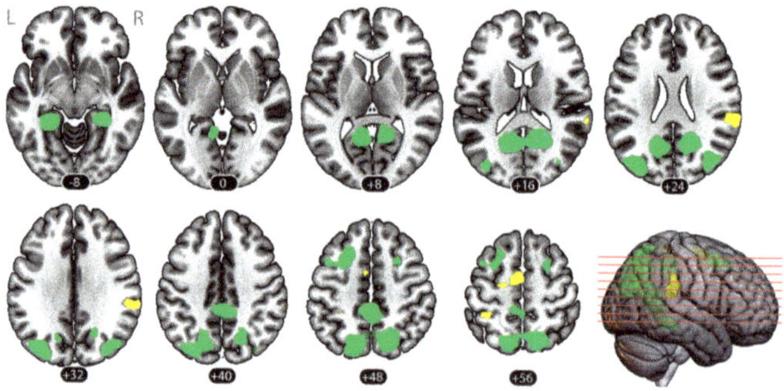

Figure 4. Specific neural correlates of the two closed-eye imagery tasks compared to each other: group map, $n = 26$, voxel-level threshold $p < 0.001$, cluster-wise threshold $p_{FDR} < 0.05$. The tennis > house navigation contrast is shown in yellow, the house navigation > tennis contrast in green. The statistical map is overlaid onto an MNI template. All data are presented in the MNI space.

Table 2. Clusters of activation specific to imaginary tennis and house navigation tasks (compared to the other task) across both vantage points in all participants.

Cluster	Volume, Voxels (mm³)	p_{FDR}	MNI Coordinates (Center of Mass) x	y	z	Region Labels [1]
Tennis > Navigation						
(1) SMA	56 (3584)	0.001	−9	−7	59	Supplementary Motor Cortex (SMA), LR Precentral Gyrus (PreCG), L
(2) Right Supramarginal Gyrus	51 (3264)	0.001	61	−34	27	Supramarginal Gyrus, posterior division (pSMG), R Supramarginal Gyrus, anterior division (aSMG), R Parietal Operculum Cortex (PO), R
(3) Left Superior Parietal	32 (2048)	0.008	−35	−43	60	Superior Parietal Lobule (SPL), L
Navigation > Tennis						
(1) Left Para-hippocampal	51 (3264)	0.003	−26	−41	−10	Parahippocampal Gyrus, posterior division (pPaHC), L Lingual Gyrus (LG), L Hippocampus L
(2) Right Para-hippocampal	36 (2304)	0.010	27	−38	−10	Parahippocampal Gyrus, posterior division (pPaHC), R Lingual Gyrus (LG), R Hippocampus R
(3) Bilateral Occipito-temporal	766 (49,024)	<0.001	−4	−62	34	Precuneous Lateral Occipital Cortex, superior division (sLOC), LR Cingulate Gyrus, posterior division (PC)
(4) Right Lateral Occipital Cortex	74 (4736)	0.001	38	−76	28	Lateral Occipital Cortex, superior division (sLOC), R
(5) Left Frontal	66 (4224)	0.001	−28	13	50	Middle Frontal Gyrus (MFG), L Superior Frontal Gyrus (SFG), L
(6) Right Frontal	22 (1408)	0.035	26	9	53	Middle Frontal Gyrus (MFG), R Superior Frontal Gyrus (SFG), R

[1] For each cluster, only the anatomical labels covering more than 5% of cluster volume are presented; voxel size is 4 mm isotropic.

Finally, the analysis revealed no significant differences in activation elicited by the first-person and the third-person vantage points in either of the imagery tasks, nor in both tasks analyzed together.

3.3. *Task-Based Functional Connectivity*

3.3.1. Voxel-to-Voxel Connectivity: Intrinsic Connectivity Contrast

Neither parametric nor nonparametric statistics revealed any significant differences in whole-brain ICC between the first-person and third-person vantage points in either of the imagery tasks, nor in both tasks considered together.

3.3.2. ROI-to-ROI Connectivity

The whole-brain atlas-wide exploratory ROI-to-ROI analysis implicated 5460 connections among 105 ROIs. For the imaginary tennis and imaginary house navigation tasks, considered individually, the NBS analysis found no differences between the first-person and third-person vantage points. However, for both tasks taken together, the NBS analysis identified a network of four ROIs and three connections showing greater functional connectivity during the first-person imagery compared to the third-person imagery, regardless of the specific task (Mass = 108.87, $p_{FDR} = 0.045$). The ROIs included the left intracalcarine cortex (ICCl), the right supracalcarine cortex (SCCr), the posterior division of the left superior temporal gyrus (pSTGl), and the posterior division of the left middle temporal gyrus (pMTGl). The details of the connections are presented in Figure 5 and Table 3.

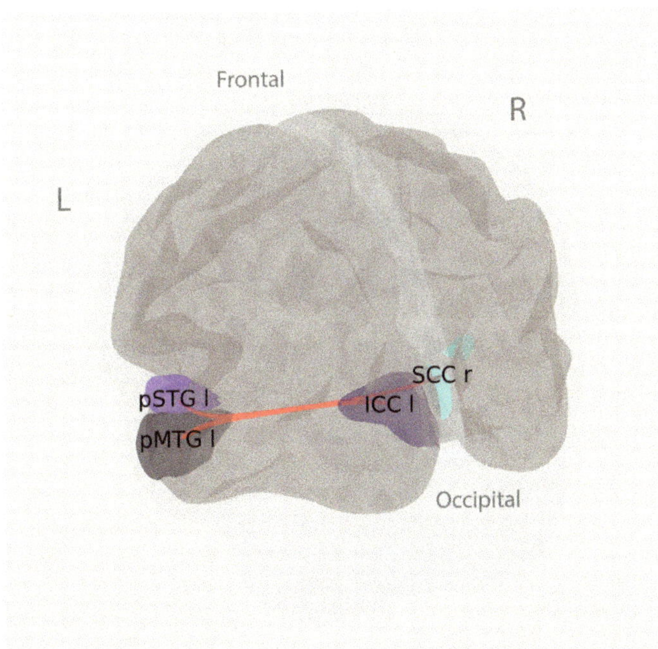

Figure 5. A network identified by the NBS connectivity analysis (mass/intensity) with an FDR correction for multiple comparisons and demonstrating greater functional connectivity in the first-person vantage point in visual imagery, compared to the third-person vantage point. ICC l: the left intracalcarine cortex; SCC r: the right supracalcarine cortex; pSTG l: the posterior division of the left superior temporal gyrus; pMTG l: the posterior division of the left middle temporal gyrus.

Table 3. Results of the NBS connectivity analysis (mass/intensity) for the first-person vs. third-person perspective contrast in both the imaginary tennis and house navigation tasks considered together.

Analysis Unit	Mass	t(24)	p-unc.	p_{FDR}
Network 1/2	108.87		0.022553	0.045
Connection ICCl—pSTGl		4.76	0.000077	0.421
Connection pMTGl—ICCl		4.19	0.000327	0.800
Connection pMTGl—SCCr		3.78	0.000926	0.999

ICCl: the intracalcarine cortex, left; pSTGl: the superior temporal gyrus, posterior division, left; pMTGl: the middle temporal gyrus, posterior division, right.

4. Discussion

4.1. Brain Activity and Connectivity Specific for the First-Person and Third-Person Perspectives in Imagery

The present study has demonstrated that different perspectives while performing mental imagery tasks such as imaginary tennis and house navigation result in an unexpectedly subtle difference in brain activity and connectivity. This is especially surprising since perspective taking seems to be a fundamental ability that underlies human thinking.

The two closed-eye imagery tasks used in our study elicited activation in the frontoparietal control network, typical of voluntary visual mental imagery, as suggested by a recent meta-analysis [41]. In line with the meta-analysis results, our data did not reveal any significant group-level activation in the early visual cortex during the closed-eye imagery tasks. On the contrary, some portions of the early visual and auditory cortices, along with the DMN components, were more activated during the baseline (resting) conditions compared to visual mental imagery, which is consistent with the idea that the early visual cortex may be involved in mental imagery through inhibitory connections [42]. The revealed task-specific activation also aligns with the task demands and the previous literature: the imaginary tennis task demonstrated greater engagement of SMA, while the imaginary navigation task demonstrated greater engagement of the parahippocampal place area (PPA; e.g., [30]).

No significant differences were found in activation elicited by the first-person and third-person conditions. This lack of difference may support the idea that at least some previous finding associating perspective taking with functioning of the DMN system (e.g., [21]) may be due to an imbalance in the difficulty of first-person and third-person perspective taking. However, the third-person perspective that demonstrated greater recruitment of the precuneus in the study by Ruby and Decety is generally associated with greater mental effort (e.g., [6,7]), while the DMN is usually more activated under the less difficult condition. Our results suggest that the ratings of the relative subjective difficulty of the two imaginary vantage points may substantially vary across individuals: more than half of our participants reported either the third-person perspective to be easier to take at least in one task, or both perspectives being equally difficult.

Such a proportion may indicate the effect of modern visual media (e.g., video games, movies, virtual reality), which frequently present experiences from a third-person perspective. Exposure to these forms of visualization may have trained individuals to adopt external perspectives, reducing the perceived cognitive effort. This increased exposure may facilitate greater flexibility in adopting different vantage points. Although such an assumption needs further empirical testing, existing evidence suggests that virtual reality facilitates seamless transitions between first- and third-person perspectives, and that users perceive perspective not as a strict dichotomy but as a continuum of experience [43]. This adaptability may reduce the cognitive cost traditionally associated with switching between perspectives [7,44], thereby minimizing the expected differences in neural and behavioral measures.

Guterstam [45] demonstrated that, under specific conditions, healthy individuals can experience both dual body ownership and dual self-location simultaneously. A similar experience may occur when participants simultaneously imagine themselves being inside the scene from a first-person perspective while also observing it from an external, third-person perspective; such simultaneous co-existence of vantage points in imagery has been hypothesized for sports imagery [46]. In the same vein, our instruction did not fully remove the participants from the imaginary scene, but rather combined the first-person view with an additional third-person perspective. The participants might still be mentally anchored in an egocentric way, merely observing from a shifted viewpoint, rather than truly stepping outside themselves. This would make third-person and first-person imagery more similar than expected, thus minimizing the differences between the recruited neural substrates.

The lack of difference between the vantage points could indicate a lack of full decentration in other tasks similar to ours, where participants are instructed to observe themselves while remaining inside the scene. This finding is particularly relevant to therapeutic approaches for post-traumatic stress disorder (PTSD), which often employ third-person perspective techniques as a way to reshape subsequent memories and to reduce emotional intensity through decentration [29]. However, our results suggest that simply instructing individuals to observe themselves from an external viewpoint may not be sufficient to fully detach them from the scene.

At the same time, our findings did not provide support for the idea that the first-person perspective loads the bodily self system more than the third-person perspective does, as we initially hypothesized. A possible explanation may arise from the fact that the model by St. Jacques (2019) [29] is based on autobiographical memory research, while our study addressed mental imagery. According to a two-dimensional theory of visual perspective, following Nigro and Neisser's original definition of the third-person perspective [3], first-person vantage point predominance is only one dimension of mental visual perspective, while the other is third-person self-visibility. Due to this two-dimensional nature, first-person and third-person perspectives may interact with the type of representational process, and therefore, the same mental vantage point would have different effects in terms of affect-ladenness, motivation, and vividness on more recent vs. older autobiographical memories, mental imagery, and episodic future thinking [47].

It is noteworthy that the only alteration of the neural correlates of visual mental imagery caused by a change in the vantage point in our study was found in the functional connectivity domain rather than activation, thus indicating that perspective taking may reshape the current processes within the mental imagery network rather than recruiting additional substrates. The first-person perspective is characterized by greater FC in the subnetwork connecting the early visual cortex with the posterior middle and superior temporal gyri in the left hemisphere. The posterior middle temporal cortex is implicated in controlled retrieval from the semantic memory [48], and therefore, the observed functional connectivity change might indicate that the first-person perspective features a closer sensory recruitment loop (in terms of working memory theory, see Phylactou et al., 2022 for a recent review [49]), or a loop between the "visual buffer" and the contents of long-term memory in terms of the quasi-pictorial theory of mental imagery [5,50]. Along with the fact that some parts of the early visual cortex in our study were deactivated during the closed-eye visual imagery tasks compared to baseline (rest), the observed FC change in its other parts may indicate that the functioning of the above-mentioned loop involves a more complex and nuanced mechanism than simply activation of the entire early visual cortex during mental image construction.

4.2. Study Limitations

Our study had several limitations. First, the Coma Science Group imagery tasks that we used do not implement any form of control of the imagery content; therefore, we have no means to ensure that the participants followed the task instructions other than their reports and anecdotal comments, e.g., complaining of not having enough space in their house to step out of their imaginary self. One direction of future research may be the development of a more controlled task that would allow the experimenters to verify the vantage-point aspect of the imager's internal experience.

Second, we used only two visual-motor imagery tasks, which constrains the possible generalization of our results.

Finally, our study may be underpowered due to its exploratory nature. We were unable to perform the power analysis and estimate the required sample size due to the lack of previous studies addressing the same effects. Not even regions of interest for potential effects were known to us in advance. Therefore, we collected a sample typical for this field of study (cf. the sample sizes reported in papers selected for the Spagna et al.) meta-analysis, which rarely exceed 20 participants [41]).

5. Conclusions

The present work shows that, despite the great importance of visual perspective taking for the development of human cognition, the difference in the brain mechanisms underlying the first-person and the third-person perspective in visual mental imagery may be subtle and is seen in the functional connectivity domain rather than activation. This fact indicates that the perspective taking modulates the functioning of the mental imagery network rather than recruiting different substrates. The observed alterations affected a small subnetwork, including some portions of the early visual cortex and the left posterior temporal areas, which manifested stronger functional connectivity during first-person perspective imagery. This finding suggests a closer sensory recruitment loop, or, in terms of the quasi-pictorial theory of mental imagery, a loop between long-term memory and the "visual buffer", for the first-person mental vantage point.

Future research should explore possible ways to better isolate third-person perspective effects as well as the effects of the two aspects of perspective taking: the ability to decenter/detach the vantage point and the ability to visualize one's own body in a mental image constructed from an external viewpoint, each of which may have their own neural correlates.

Author Contributions: E.P., conceptualization, methodology, formal analysis, funding acquisition, writing—original draft preparation; M.R., investigation, formal analysis, writing—review and editing; V.V., project administration, investigation, writing—review and editing; O.B., conceptualization, writing—review and editing; E.M., methodology, investigation, data curation, resources, writing—review and editing. All authors have read and agreed to the published version of the manuscript.

Funding: This research was supported by the Russian Science Foundation, grant #23-28-01561, https://rscf.ru/project/23-28-01561/ (accessed on 1 March 2025); Olesya Blazhenkova was supported by Personal Research Fund grant from Sabancı University.

Institutional Review Board Statement: The study was conducted in accordance with the Declaration of Helsinki, and approved by the Local Ethics Committee of the Medical Research and Educational Centre, Lomonosov Moscow State University (protocol 04/22, 16 May 2022).

Informed Consent Statement: Informed consent was obtained from all subjects involved in the study.

Data Availability Statement: The data reported in the paper are available in the Openneuro repository: https://doi.org/10.18112/openneuro.ds006077.v1.0.0 (accessed on 2 April 2025).

Acknowledgments: We are grateful to all participants of our experiment and thank Semen Mening, Anna Smirnova, Maria Pchelintseva, and Vera Ivanushko for technical assistance.

Conflicts of Interest: The authors declare no conflicts of interest.

References

1. Tversky, B.; Hard, B.M. Embodied and Disembodied Cognition: Spatial Perspective-Taking. *Cognition* **2009**, *110*, 124–129. [CrossRef] [PubMed]
2. Piaget, J. *The Construction of Reality in the Child*; Basic Books: New York, NY, USA, 1954.
3. Nigro, G.; Neisser, U. Point of View in Personal Memories. *Cogn. Psychol.* **1983**, *15*, 467–482. [CrossRef]
4. Libby, L.K.; Eibach, R.P. Visual Perspective in Mental Imagery. In *Advances in Experimental Social Psychology*; Elsevier: Amsterdam, The Netherlands, 2011; Volume 44, pp. 185–245. [CrossRef]
5. Kosslyn, S.M.; Ganis, G.; Thompson, W.L. Neural Foundations of Imagery. *Nat. Rev. Neurosci.* **2001**, *2*, 635–642. [CrossRef]
6. Zacks, J.M.; Mires, J.; Tversky, B.; Hazeltine, E. Mental spatial transformations of objects and perspective. *Spat. Cogn. Comput.* **2000**, *2*, 315–332. [CrossRef]
7. Hong, J.P. The Influence of Visual Perspective on the Cognitive Effort Required for Mental Representation. Ph.D. Thesis, Wilfrid Laurier University, Waterloo, ON, Canada, 2024.
8. Hegarty, M. A Dissociation between Mental Rotation and Perspective-Taking Spatial Abilities. *Intelligence* **2004**, *32*, 175–191. [CrossRef]
9. Galvan Debarba, H.; Bovet, S.; Salomon, R.; Blanke, O.; Herbelin, B.; Boulic, R. Characterizing First and Third Person Viewpoints and Their Alternation for Embodied Interaction in Virtual Reality. *PLoS ONE* **2017**, *12*, e0190109. [CrossRef]
10. Jackson, P.L.; Meltzoff, A.N.; Decety, J. Neural Circuits Involved in Imitation and Perspective-Taking. *NeuroImage* **2006**, *31*, 429–439. [CrossRef]
11. Higuchi, T.; Nagami, T.; Nakata, H.; Watanabe, M.; Isaka, T.; Kanosue, K. Contribution of Visual Information about Ball Trajectory to Baseball Hitting Accuracy. *PLoS ONE* **2016**, *11*, e0148498. [CrossRef]
12. Avraamides, M.N.; Klatzky, R.L.; Loomis, J.M.; Golledge, R.G. Use of Cognitive Versus Perceptual Heading During Imagined Locomotion Depends on the Response Mode. *Psychol. Sci.* **2004**, *15*, 403–408. [CrossRef]
13. Burgess, N. Spatial Memory: How Egocentric and Allocentric Combine. *Trends Cogn. Sci.* **2006**, *10*, 551–557. [CrossRef]
14. Klatzky, R.L. Allocentric and Egocentric Spatial Representations: Definitions, Distinctions, and Interconnections. In *Spatial Cognition*; Freksa, C., Habel, C., Wender, K.F., Eds.; Lecture Notes in Computer Science; Springer: Berlin/Heidelberg, Germany, 1998; Volume 1404, pp. 1–17. [CrossRef]
15. Zaehle, T.; Jordan, K.; Wüstenberg, T.; Baudewig, J.; Dechent, P.; Mast, F.W. The Neural Basis of the Egocentric and Allocentric Spatial Frame of Reference. *Brain Res.* **2007**, *1137*, 92–103. [CrossRef] [PubMed]
16. Kozhevnikov, M.; Hegarty, M. A Dissociation between Object Manipulation Spatial Ability and Spatial Orientation Ability. *Mem. Cogn.* **2001**, *29*, 745–756. [CrossRef]
17. Vogeley, K.; Fink, G.R. Neural Correlates of the First-Person-Perspective. *Trends Cogn. Sci.* **2003**, *7*, 38–42. [CrossRef] [PubMed]
18. Leng, X.; Zhu, W.; Mayer, R.E.; Wang, F. The Viewing Perspective Effect in Learning from Instructional Videos: A Replication and Neuroimaging Extension. *Learn. Instr.* **2024**, *94*, 102004. [CrossRef]
19. Pavone, E.F.; Tieri, G.; Rizza, G.; Tidoni, E.; Grisoni, L.; Aglioti, S.M. Embodying Others in Immersive Virtual Reality: Electro-Cortical Signatures of Monitoring the Errors in the Actions of an Avatar Seen from a First-Person Perspective. *J. Neurosci.* **2016**, *36*, 268–279. [CrossRef]
20. Cammisuli, D.M.; Castelnuovo, G. Neuroscience-Based Psychotherapy: A Position Paper. *Front. Psychol.* **2023**, *14*, 1101044. [CrossRef]
21. Ruby, P.; Decety, J. Effect of Subjective Perspective Taking during Simulation of Action: A PET Investigation of Agency. *Nat. Neurosci.* **2001**, *4*, 546–550. [CrossRef]
22. Tomasino, B.; Werner, C.J.; Weiss, P.H.; Fink, G.R. Stimulus Properties Matter More than Perspective: An fMRI Study of Mental Imagery and Silent Reading of Action Phrases. *NeuroImage* **2007**, *36*, T128–T141. [CrossRef]
23. Arzy, S.; Thut, G.; Mohr, C.; Michel, C.M.; Blanke, O. Neural Basis of Embodiment: Distinct Contributions of Temporoparietal Junction and Extrastriate Body Area. *J. Neurosci.* **2006**, *26*, 8074–8081. [CrossRef]
24. Blanke, O.; Mohr, C.; Michel, C.M.; Pascual-Leone, A.; Brugger, P.; Seeck, M.; Landis, T.; Thut, G. Linking Out-of-Body Experience and Self Processing to Mental Own-Body Imagery at the Temporoparietal Junction. *J. Neurosci.* **2005**, *25*, 550–557. [CrossRef]
25. Smith, A.M.; Messier, C. Voluntary Out-of-Body Experience: An fMRI Study. *Front. Hum. Neurosci.* **2014**, *8*, 70. [CrossRef]
26. Gauthier, B.; Bréchet, L.; Lance, F.; Mange, R.; Herbelin, B.; Faivre, N.; Bolton, T.A.W.; Ville, D.V.D.; Blanke, O. First-Person Body View Modulates the Neural Substrates of Episodic Memory and Autonoetic Consciousness: A Functional Connectivity Study. *NeuroImage* **2020**, *223*, 117370. [CrossRef] [PubMed]

27. Grol, M.; Vingerhoets, G.; De Raedt, R. Mental Imagery of Positive and Neutral Memories: A fMRI Study Comparing Field Perspective Imagery to Observer Perspective Imagery. *Brain Cogn.* **2017**, *111*, 13–24. [CrossRef] [PubMed]
28. Eich, E.; Nelson, A.L.; Leghari, M.A.; Handy, T.C. Neural Systems Mediating Field and Observer Memories. *Neuropsychologia* **2009**, *47*, 2239–2251. [CrossRef] [PubMed]
29. St. Jacques, P.L. A New Perspective on Visual Perspective in Memory. *Curr. Dir. Psychol. Sci.* **2019**, *28*, 450–455. [CrossRef]
30. Owen, A.M.; Coleman, M.R.; Boly, M.; Davis, M.H.; Laureys, S.; Pickard, J.D. Detecting Awareness in the Vegetative State. *Science* **2006**, *313*, 1402. [CrossRef]
31. Oldfield, R.C. The Assessment and Analysis of Handedness: The Edinburgh Inventory. *Neuropsychologia* **1971**, *9*, 97–113. [CrossRef]
32. Fernández-Espejo, D.; Norton, L.; Owen, A.M. The Clinical Utility of fMRI for Identifying Covert Awareness in the Vegetative State: A Comparison of Sensitivity between 3T and 1.5T. *PLoS ONE* **2014**, *9*, e95082. [CrossRef]
33. Desikan, R.S.; Ségonne, F.; Fischl, B.; Quinn, B.T.; Dickerson, B.C.; Blacker, D.; Buckner, R.L.; Dale, A.M.; Maguire, R.P.; Hyman, B.T.; et al. An Automated Labeling System for Subdividing the Human Cerebral Cortex on MRI Scans into Gyral Based Regions of Interest. *NeuroImage* **2006**, *31*, 968–980. [CrossRef]
34. Nieto-Castanon, A. *Handbook of Functional Connectivity Magnetic Resonance Imaging Methods in CONN*; Hilbert Press: Boston, MA, USA, 2020; ISBN 978-0-578-64400-4.
35. Whitfield-Gabrieli, S.; Nieto-Castanon, A.; Ghosh, S. *Artifact Detection Tools (ART)*; Release Version 7:11; MIT: Cambridge, MA, USA, 2011.
36. Behzadi, Y.; Restom, K.; Liau, J.; Liu, T.T. A Component Based Noise Correction Method (CompCor) for BOLD and Perfusion Based fMRI. *NeuroImage* **2007**, *37*, 90–101. [CrossRef]
37. Martuzzi, R.; Ramani, R.; Qiu, M.; Shen, X.; Papademetris, X.; Constable, R.T. A Whole-Brain Voxel Based Measure of Intrinsic Connectivity Contrast Reveals Local Changes in Tissue Connectivity with Anesthetic without a Priori Assumptions on Thresholds or Regions of Interest. *NeuroImage* **2011**, *58*, 1044–1050. [CrossRef] [PubMed]
38. Chumbley, J.; Worsley, K.; Flandin, G.; Friston, K. Topological FDR for Neuroimaging. *NeuroImage* **2010**, *49*, 3057–3064. [CrossRef]
39. Bullmore, E.T.; Suckling, J.; Overmeyer, S.; Rabe-Hesketh, S.; Taylor, E.; Brammer, M.J. Global, Voxel, and Cluster Tests, by Theory and Permutation, for a Difference between Two Groups of Structural MR Images of the Brain. *IEEE Trans. Med. Imaging* **1999**, *18*, 32–42. [CrossRef]
40. Zalesky, A.; Fornito, A.; Bullmore, E.T. Network-Based Statistic: Identifying Differences in Brain Networks. *NeuroImage* **2010**, *53*, 1197–1207. [CrossRef] [PubMed]
41. Spagna, A.; Hajhajate, D.; Liu, J.; Bartolomeo, P. Visual Mental Imagery Engages the Left Fusiform Gyrus, but Not the Early Visual Cortex: A Meta-Analysis of Neuroimaging Evidence. *Neurosci. Biobehav. Rev.* **2021**, *122*, 201–217. [CrossRef] [PubMed]
42. Dijkstra, N. Uncovering the Role of the Early Visual Cortex in Visual Mental Imagery. *Vision* **2024**, *8*, 29. [CrossRef]
43. Hoppe, M.; Baumann, A.; Tamunjoh, P.C.; Machulla, T.-K.; Woźniak, P.W.; Schmidt, A.; Welsch, R. There Is No First- or Third-Person View in Virtual Reality: Understanding the Perspective Continuum. In Proceedings of the CHI Conference on Human Factors in Computing Systems, New Orleans, LA, USA, 29 April–5 May 2022; ACM: New York, NY, USA, 2022; pp. 1–13.
44. Aïte, A.; Berthoz, A.; Vidal, J.; Roëll, M.; Zaoui, M.; Houdé, O.; Borst, G. Taking a Third-Person Perspective Requires Inhibitory Control: Evidence from a Developmental Negative Priming Study. *Child Dev.* **2016**, *87*, 1825–1840. [CrossRef]
45. Guterstam, A.; Larsson, D.E.O.; Szczotka, J.; Ehrsson, H.H. Duplication of the Bodily Self: A Perceptual Illusion of Dual Full-Body Ownership and Dual Self-Location. *R. Soc. Open Sci.* **2020**, *7*, 201911. [CrossRef]
46. Sutton, J. Memory Before the Game: Switching Perspectives in Imagining and Remembering Sport and Movement. *J. Ment. Imag.* **2012**, *36*, 85–95.
47. Kinley, I.; Porteous, M.; Levy, Y.; Becker, S. Visual Perspective as a Two-Dimensional Construct in Episodic Future Thought. *Conscious. Cogn.* **2021**, *93*, 103148. [CrossRef]
48. Davey, J.; Thompson, H.E.; Hallam, G.; Karapanagiotidis, T.; Murphy, C.; De Caso, I.; Krieger-Redwood, K.; Bernhardt, B.C.; Smallwood, J.; Jefferies, E. Exploring the Role of the Posterior Middle Temporal Gyrus in Semantic Cognition: Integration of Anterior Temporal Lobe with Executive Processes. *NeuroImage* **2016**, *137*, 165–177. [CrossRef] [PubMed]
49. Phylactou, P.; Traikapi, A.; Papadatou-Pastou, M.; Konstantinou, N. Sensory Recruitment in Visual Short-Term Memory: A Systematic Review and Meta-Analysis of Sensory Visual Cortex Interference Using Transcranial Magnetic Stimulation. *Psychon. Bull. Rev.* **2022**, *29*, 1594–1624. [CrossRef] [PubMed]
50. Kosslyn, S.M. *Image and Mind*; Harvard University Press: Cambridge, MA, USA; London, UK, 1980; ISBN 978-0-674-44366-2.

vision

Article

The Phenomenology of Offline Perception: Multisensory Profiles of Voluntary Mental Imagery and Dream Imagery

Maren Bilzer [1] and Merlin Monzel [1,2,*]

[1] Department of Psychology, Personality Psychology and Biological Psychology, University of Bonn, Kaiser-Karl-Ring 9, 53111 Bonn, Germany; maren.bilzer@uni-bonn.de
[2] Centre for Clinical Brain Sciences, University of Edinburgh, Edinburgh EH16 4SB, UK
* Correspondence: merlin.monzel@uni-bonn-diff.de

Abstract: Both voluntary mental imagery and dream imagery involve multisensory representations without externally present stimuli that can be categorized as offline perceptions. Due to common mechanisms, correlations between multisensory dream imagery profiles and multisensory voluntary mental imagery profiles were hypothesized. In a sample of 226 participants, correlations within the respective state of consciousness were significantly bigger than across, favouring two distinct networks. However, the association between the vividness of voluntary mental imagery and vividness of dream imagery was moderated by the frequency of dream recall and lucid dreaming, suggesting that both networks become increasingly similar when higher metacognition is involved. Additionally, the vividness of emotional and visual imagery was significantly higher for dream imagery than for voluntary mental imagery, reflecting the immersive nature of dreams and the continuity of visual dominance while being awake and asleep. In contrast, the vividness of auditory, olfactory, gustatory, and tactile imagery was higher for voluntary mental imagery, probably due to higher cognitive control while being awake. Most results were replicated four weeks later, weakening the notion of state influences. Overall, our results indicate similarities between dream imagery and voluntary mental imagery that justify a common classification as offline perception, but also highlight important differences.

Keywords: offline perception; phantom perception; voluntary mental imagery; dream imagery; multisensory imagery; dream recall; lucid dreaming

1. Introduction

"Dreams, they feel real while we are in them, right? It's only when we wake up that we realize that something was actually strange".

In Christopher Nolan's science fiction film *Inception*, the protagonist Dominick Cobb raises a central philosophical question that is also discussed in research [1]: How can we ever be sure that we are awake and not dreaming? Dreaming is often perceived as an act of imagination [2,3], because, like imagining, it creates simulations of perception. Thus, the question can be extended: How can we ever be sure that we are actually perceiving external stimuli and not just imagining? Therefore, the present study aims to investigate the extent to which dream imagery and voluntary mental imagery are phenomenologically distinguishable and how they differ from sensory perceptions in terms of vividness by correlating the vividness ratings of the different sensory modalities for dream imagery and voluntary mental imagery, considering the roles of dream memory and lucid dreaming as moderators. Within this context, vividness refers to the measure of how close the mental image comes to an actual perception [4].

1.1. Dream Imagery and Voluntary Mental Imagery as Offline Perceptions

Dream imagery and voluntary mental imagery can both be understood as sensory representations in the absence of external stimuli [5–7]. Thus, they are often commonly classified as phantom [8] or offline perceptions [9]. An important difference is that voluntary mental imagery is generated voluntarily, whereas dreams mostly lack volitional control [5,7,8]. Neuronally, both types of imagery share the default mode network [5,10], and people who frequently recall their dreams have stronger activations in the default mode network while dreaming and being awake [11]. In addition, the loss of voluntary visual imagery may coincide with the loss of visual dream imagery as a result of damage to the hippocampus and disrupted visual pathway connections [12].

Voluntary mental imagery is thought to follow the same neural pathways as perception, albeit top-down. Accordingly, imagining functions like perception in reverse and starts a cascade beginning in frontal areas to the visual cortex [5]. Evidence for the assumption of voluntary mental imagery being a weaker form of perception is provided by visual imagery priming [13]. For example, ref. [14] demonstrated that imagining a stimulus influences which stimulus enters consciousness in a binocular rivalry task. Ref. [15] showed a priming effect of imagining colours on the perception of colours, and [16] as well as [17] found that mental imagery primes a visual search. Another study indicated that imagery resembles the characteristics of perceptions by showing that participants' pupils dilated or constricted corresponding to the brightness of the imagined stimulus [18]. Moreover, visual mental imagery can be decoded using fMRI activity patterns in the visual cortices [19]. Ref. [20] showed that perceptual learning was possible through mental imagery alone. Furthermore, it was shown that people who were born blind were impaired in mental imagery [21]. Last, ref. [22] revealed that higher ratings of voluntary mental imagery vividness were associated with a smaller difference between perception and imagery in the activity patterns of visual areas.

The communalities between dreams and perception can also be elaborated. Ref. [23] applied the decoding of visual dream content using machine-learning models trained on stimulus-induced fMRI activity as a neural measure of the visual dream content. Images were assigned to specific fMRI activities based on activity during perception. The accuracy of decoding was confirmed by dream reports. Ref. [24] found that anterior cortical regions were activated during thought-like dream experiences and posterior cortical regions were activated during perceptual dream experiences with vivid sensory contents. Faces, for example, were accompanied by activation in the fusiform face area, and dreams with a specific spatial environment were associated with increased activity in the right posterior parietal cortex. Ref. [25] proposed that our eyes scan the dream scene during REM sleep, providing further evidence for the visual properties of dreams. Overall, these results indicate that dream imagery and voluntary mental imagery show many neural and behavioural commonalities with perception. Derived from this, we will examine the phenomenological overlap with regard to multisensory experiences.

1.2. Multisensory Profiles of Voluntary Mental Imagery and Dream Imagery

Voluntary mental imagery is most frequently investigated within the visual domain [26]. However, mental imagery is multisensory in nature and includes additional modalities, such as auditory, olfactory, gustatory, tactile, proprioceptive, and emotional imagery [27]. The vividness of mental imagery differs depending on the respective sensory system. Vividness is rated highest for visual impressions, followed by tactile and auditory impressions, whereas the ratings for emotional and proprioceptive impressions and especially for gustatory and olfactory impressions are lower [28]. The ratings vary to some extent; for example, ref. [29] found that mental imagery especially involves auditory imagery, but also that auditory and visual imagery occur more often in both imagination and dreams than gustatory, olfactory, and tactile imagery. Importantly, judgements of vividness correlate with the activation in associated modality-specific brain regions [30,31]. A left-lateralised modality-general network as well as modality-specific sensory areas could be identified [6,32].

Dream imagery is also multisensory in nature. Ref. [33] gives an overview of various studies of the predominant sensory modalities in dreams. Visual impressions were most common (100%), followed by auditory impressions (65%). In contrast, only 1% reported experiencing gustatory, olfactory, or tactile sensations [34]. A review [35] presents predominantly visual imagery, followed by frequent auditory impressions and less frequent gustatory, olfactory, vestibular, and tactile impressions. A study by [36] found less frequent reports of gustatory, olfactory, and tactile experiences than reports of visual, auditory, and kinaesthetic modalities. The subjects were thus divided into three groups: firstly, participants who reported experiencing mainly visual, auditory, and kinaesthetic imagery; secondly, a group who experienced multisensory imagery; and thirdly, a group who reported no sensory experience at all. The group with multisensory dream imagery also showed the highest ability to generate voluntary mental imagery while being awake, suggesting a common cognitive mechanism. Importantly, many of the aforementioned studies measured the frequency of various sensory experiences in dream imagery, whereas the vividness of dream imagery was less examined. We aim to explore the vividness of dream imagery to make it more comparable to the phenomenology of voluntary imagery.

1.3. The Continuity Hypothesis

The *continuity hypothesis* postulates a continuity between experiences while being awake and dreaming [37]. Studies indicate a continuity between the same conceptions and personal concerns, as well as between activities during dreaming and being awake [38]. Negative and positive daytime emotions influence dream emotions and vice versa [39]. Additionally, ref. [40] found similarities in cognitive and sensory qualities, suggesting the same cognitive–perceptual system being active during both experiences. Recent neurophysiological perspectives on the continuity hypothesis indicated similar underlying neuronal mechanisms and the same cognitive and emotional functioning across different states of consciousness [41].

A similar continuity hypothesis can be proposed for between perception and voluntary mental imagery. The *constructive episodic simulation hypothesis* (CESH) [42,43] postulates that remembering and imagining share the same characteristics and depend on constructive and simulative processes. When we remember a past experience, we access different independent elements from our memory to reconstruct a perceptual representation. When imaging something, a new representation is constructed from the same data base we use for remembering. Thus, our imagination is dependent on previous perceptions. Ref. [44] expands the CESH to the CESH+, including visual, auditory, gustatory, tactile, olfactory, and emotional sensory modalities. The same sensory brain areas are activated while generating episodic memory and voluntary mental imagery [6,45]. However, while remembering and imagining are mostly voluntary processes (excluding phenomena such as daydreaming, mind wandering, and intrusive memory), dream imagery is involuntary. Ref. [44] explains that dreaming also includes this constructive process but the elements to construct dream imagery can be retrieved without intention, referring to evidence provided by [46], who found that the dreams of patients with hippocampal damage include fewer episodic details, such as multisensory imagery.

Overall, it seems apparent that our sensory perceptions shape our voluntary sensory imagery as well as our sensory dream imagery in terms of memory. Therefore, it is conceivable that people who experience more sensations in one modality (e.g., a perfumer) develop both more voluntary mental imagery *and* dream imagery within this modality.

Hypothesis 1. *There are significant positive correlations between the vividness ratings of different sensory modalities in voluntary mental imagery and the same sensory modalities in dream imagery.*

1.4. Dissociating Dream Imagery and Voluntary Mental Imagery

The conception of a common mechanism between voluntary mental imagery and dream imagery is challenged by the phenomenon of aphantasia, defined as a reduced or absent ability to produce voluntary mental imagery. Around 26.22% of people with aphantasia report a complete absence of multisensory imagery [47] while still being able to produce dream imagery [48]. However, they dream significantly less frequently and less vividly [47]. The fact that aphantasics can experience dreams may be reflected in a disturbance higher in the cascade of imagery generation rather than in sensory cortices [49]. Therefore, although sharing some neural substrates, dream imagery and voluntary mental imagery also rely on distinct mechanisms, leading to the following hypothesis:

Hypothesis 2. *There are stronger positive correlations of different sensory modalities within voluntary mental imagery and within dream imagery (=monodomain correlations) than across voluntary mental imagery and dream imagery (=multidomain correlations).*

Moreover, the studies presented above provided some evidence that certain sensory modalities, such as tactile, gustatory and olfactory imagery, occur very rarely or not at all in dreams. From the rare occurrence of these sensory modalities we would also assume lower vividness, since the vividness of mental imagery and frequent occurrence of mental imagery is often associated, at least while being awake, e.g., ref. [50]. For dreams, there is an association between dreams recall frequency and vividness of dream imagery [51]. Thus, we propose the following hypothesis:

Hypothesis 3. *The vividness of tactile, gustatory, and olfactory imagery is lower for dream imagery than for voluntary mental imagery.*

A last difference between dream imagery and voluntary mental imagery is the "here and now" quality and immersive nature of dreams [1,52], leading to the appearance of reality until waking up [53]. According to [54], reality testing is missing while dreaming, since frontal regions are less activated, resulting in impaired executive functions, less top-down control, and deficient reality monitoring [7]. In contrast, imagination involves frontal and parietal regions to a higher extent [55,56], which enables cognitive control and reality monitoring [57]. This becomes especially apparent when looking at the affective component of dreams. Ref. [58] found that 74% of dreams contained emotions. While voluntary mental imagery also contains emotional representations [59], dreams are marked by immersion and hallucinatory qualities, amplifying the emotions and making them more intense [60]. Thus, one would not confuse a voluntarily imagined emotion with a real emotion, unlike in dreams.

Hypothesis 4. *The vividness of emotional imagery is higher for dream imagery than for voluntary mental imagery.*

1.5. Further Influences

Although we argued for some differences between the vividness of voluntary mental imagery and the vividness of dream imagery, we also assumed some moderating factors that may attenuate the differences between both phenomena.

1.5.1. Dream Recall

Despite experiencing sensory impressions, emotions, and thoughts during dreaming, we often do not remember them after waking up. Therefore, unsuccessful dream recall restricts the access to the phenomenology of dreams. The ability to recall a dream is therefore an important prerequisite for dream research and the only access to subjective dream experiences [61].

Dream recall frequency varies between and within individuals. On the one hand, trait factors were identified, such as interest in dreams, attitude towards dreams, personality factors, and socio-demographic variables. On the other hand, state factors were identified, such as poor sleep quality, frequent night waking, the presence of a mental illness, and stress [61,62]. Another reason for forgetting dreams can be a low dream quality [63]. According to [64], a higher capacity to produce visual imagery enriches the dream experience and promotes more frequent dream recall, which could also be demonstrated in a correlation between dream recall frequency and vividness of visual and tactile voluntary imagery [51]. However, it is not entirely clear whether people actually remember more vivid dreams or whether remembered dreams only appear more vivid due to more conscious processing.

Overall, since people lack access to the vividness of their dream imagery in the absence of dream recall, dream recall is most likely a moderator of the association between vividness of dream imagery and vividness of voluntary mental imagery. By considering dream recall as a moderator, the differences between the vividness of various sensory modalities in dream imagery and voluntary mental imagery should be reduced.

Hypothesis 5. *The association between the vividness of voluntary mental imagery and vividness of dream imagery is moderated by the dream recall frequency.*

1.5.2. Lucid Dreaming

In the same vein, lucid dreaming can be perceived as a moderating factor for differences between the vividness of voluntary mental imagery and the vividness of dream imagery. Typically, dreams are not accessible to metacognition. The exception are lucid dreams, in which dreamers gain the ability to reflect on their current state of dreaming [65]. At the same time, the dreamer remains in the physiological state of sleep. In an EEG study, ref. [66] showed increased activation during lucid dreaming in the right dorsolateral prefrontal cortex, the parietal lobe, bilateral frontopolar regions, and the precuneus, that is, in regions associated with metacognition; higher cognitive capacities; processing of one's own thoughts, feelings, and actions; as well as experiences from the first-person perspective [67–69]. In contrast, these regions exhibit lower activity in non-lucid dreams [70]. Ref. [71] concludes that lucid dreaming is not a pure REM sleep phenomenon, but has characteristics of both REM sleep and the waking state and is thus to be understood as a hybrid state of consciousness. By understanding lucid dreams as a hybrid state, it can be argued for connections between the vividness of voluntary mental imagery and dream imagery. For example, ref. [72] demonstrated that the vividness of multisensory imagery during wakefulness predicts the frequency of lucid dreaming and [73] showed that lucid dreams are rated as more visual, auditory, and emotional than non-lucid dreams.

Overall, since dreamers gain executive control during lucid dreaming, lucid dreams are processed more deeply and experiences can be better remembered and reported, leading to a stronger association between the vividness of voluntary mental imagery and dream imagery.

Hypothesis 6. *The association between the vividness of voluntary mental imagery and the vividness of dream imagery is moderated by the frequency of lucid dreams.*

1.5.3. State Effects

Last, state effects on vividness ratings must be considered. Ref. [74] states that mental imagery vividness should not be understood as a stable maximum performance trait or an ability but as a dynamic performance trait. Accordingly, retest reliabilities are limited but still too high to define mental imagery vividness as an unstable state. Potential effects of variables like social desirability, self-confidence, training, and context on imagery ratings were considered. For instance, ref. [75] showed that contextual factors and movement during imagery had an enhancing effect on vividness ratings. By understanding imagery as a dynamic process, it becomes clear that studying the phenomenology of mental imagery requires multiple testing. Furthermore, it can be assumed that the vividness of dream imagery is also dynamic rather than stable and varies from dream to dream. For example, ref. [76] found several factors affecting the dream experience, such as the emotional intensity of current life events. Therefore, we tested our participants at two different measurement times to ensure that our effects were stable and not dependent on one specific point in time.

2. Materials and Methods

2.1. Participants

The initial sample size consisted of $N = 226$ participants who completed measurement 1. Of those, $N = 158$ participants completed measurement 2. Participants were recruited by advertisement on university websites, social media, and word-of-mouth advertising. Inclusion criteria were legal age and reachability via e-mail. Among the sample, 3 participants were classified as aphantasics (VVIQ < 23; according to [48]) and 9 were classified as hyperphantasics (VVIQ > 75; [48]). The remaining participants reached a mean score on the VVIQ of $M = 57.12$ ($SD = 10.55$). The VVIQ sum scores followed a normal distribution (see Figure 1). The age ranged from 18 to 59 years ($M = 25.51$; $SD = 8.46$). The sample was composed of 186 females (82.3%), 39 males (17.3%) and 1 participant of another gender (0.4%). The level of education was generally high (7% secondary school, 70.4% A-levels, 21.2% university degree, 0.4% doctoral degree, and 0.9% other).

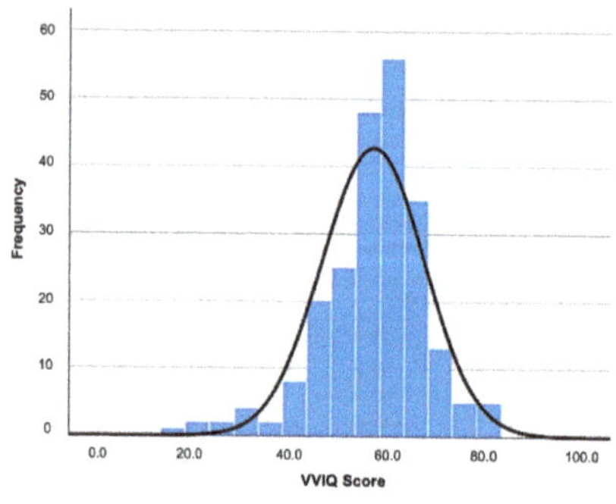

Figure 1. Distribution of VVIQ sum scores in the sample ($N = 226$). A normal distribution is superimposed over the empirical distribution.

2.2. Materials

The *Vividness of Visual Imagery Questionnaire* (VVIQ) [4] was used to measure visual voluntary mental imagery. Four different visual scenarios are described and the participants are instructed to produce the corresponding mental images. They are then asked to rate how vivid these are, that is, how closely they resemble their actual perception. There are 16 items, which are to be rated on a five-point Likert scale from 0 ("*No image at all, you only know that you are thinking of the object*") to 5 ("*Perfectly clear and as vivid as normal vision*"). An example for an item is "Visualize a rising sun. Consider carefully the picture that comes before your mind's eye. A rainbow appears".

The vividness of multisensory voluntary mental imagery was measured by means of the *Plymouth Sensory Imagery Questionnaire* (PSI-Q) [28]. The PSI-Q includes items for the following sensory modalities: vision, sound, smell, taste, touch, bodily sensation, and emotional feeling. The questionnaire consists of seven sets of five items, each regarding one modality. The participants were instructed to rate how similar their mental imagery was to actual perception. An example for an item is "Imagine the sound of an ambulance siren", which had to be rated on a 11-point scale anchored by 0 (*"No image at all"*) and 10 (*"As vivid as in real life"*).

To measure the vividness of multisensory imagery in the state of dreaming, the PSI-Q was adapted. The new form of the PSI-Q included one item for each modality concerning dreams. The instruction was to rate how similar the sensory modalities occurring in the participant's latest dreams were to actual perception on a 11-point scale from 0 (*"No image at all"*) to 10 (*"As vivid as in real life"*).

In order to examine additional features of dreaming, the *Mannheim Dream Questionnaire* (MADRE) [77] was applied. A seven-point scale was used to elicit the dream recall frequency (0 = *"never"*, 1 = *"less than once a month"*, 2 = *"about once a month"*, 3 = *"about 2 to 3 times a month"*, 4 = *"about once a week"*, 5 = *"several times a week"*, 6 = *"almost every morning"*). The lucid dreaming frequency was measured on an eight-point scale (0 = *"never"*, 1 = *"less than once a year"*, 2 = *"about once a year"*, 3 = *"about two to four times a year"*, 4 = *"about once a month"*, 5 = *"about two to three times a month"*, 6 = *"about once a week"*, 7 = *"several times a week"*), accompanied by a definition of lucid dreaming. The poles of the MADRE were reversed before the analyses to obtain an understandable interpretation. Otherwise, for example, a low score would have indicated frequent dream recall. As the MADRE does not have metric scale levels, the following three groups were formed for moderation analyses, *less than once a month*, *at least once a month* and at *least once a week*, to ensure sufficiently large subgroups.

2.3. Procedure

The study was conducted online via the software *SoSci Survey* (Version 3.2.50) [78]. The questionnaire was suitable for use with laptops, tablets, as well as smartphones. An incentive to participate was given by offering feedback regarding the ability to create multisensory imagery based on the PSI-Q and by offering student credit (0.75 subject hours) for psychology students. The study contained two measurements taken four weeks apart. The participants were automatically reminded by e-mail to take part in the second measurement. The participation in each measurement took around 10 to 15 min. Measurement 1 included the VVIQ, the original PSI-Q, the dream PSI-Q, and the three items of the MADRE described above. Measurement 2 included the VVIQ, the original PSI-Q, and the dream PSI-Q. Before the survey started, participants were informed about anonymity, voluntariness, and had to provide informed consent, followed by questions regarding sociodemographic data as gender, age, and education.

2.4. Statistical Analyses

Unless otherwise stated, analyses were conducted using SPSS (Version 27.0) [79]. Retest reliabilities and intercorrelations between the vividness of different sensory modalities within and between voluntary mental imagery and dream imagery were assessed via Pearson's correlations and presented in a heat map reflecting a multisense–multidomain matrix following conventional multitrait–multimethod matrices. The term "sense" is used to denote the sensory modality and "domain" refers to voluntary mental imagery or dream imagery. For correlation comparisons, correlations were mean-aggregated into monosense–monodomain correlations (i.e., averaging retest correlations within voluntary mental imagery and within dream imagery), monosense–multidomain correlations (i.e., averaging correlations of single sensory modalities across voluntary mental imagery and dream imagery), multisense–monodomain correlations (i.e., averaging correlations across sensory modalities within voluntary mental imagery and within dream imagery), and multisense–multidomain correlations (i.e., averaging correlations across sensory modalities across voluntary mental imagery and dream imagery) using Fisher's z-transformations, before being compared with each other using the online tool *Psychometrica* [80]. An additional network analysis was performed using JASP 0.16.2 [81] to assess the sparsity of the total model, as well as for voluntary mental imagery and dream imagery separately. A sparse network has fewer links than a denser network. Thus, a higher sparsity for the total model than for both individual models would indicate two separate networks. Additional correlation comparisons between dream recall and lucid dreaming subgroups were calculated to assess moderation effects. Paired *t*-tests were calculated to assess differences in the vividness of multisensory imagery between voluntary mental imagery and dream imagery. To cross-validate our results, all analyses were performed separately for both measurement times. Due to the larger sample size, only the analyses for measurement time 1 are reported in the main text (with the exception of the intercorrelations in Figure 2 to assess retest reliabilities), whereas the analyses for measurement time 2 can be found in the Supplementary Materials. Deviations between the two measurement times are highlighted.

		Voluntary mental imagery							Dream imagery						
		1. Vision	2. Sound	3. Smell	4. Taste	5. Touch	6. Body	7. Emotion	8. Vision	9. Sound	10. Smell	11. Taste	12. Touch	13. Body	14. Emotion
Voluntary mental imagery	1. Vision	0.66	0.7	0.55	0.56	0.57	0.6	0.48	0.37	0.23	0.23	0.19	0.3	0.25	0.3
	2. Sound	0.65	0.68	0.72	0.68	0.68	0.65	0.5	0.3	0.27	0.25	0.24	0.33	0.28	0.29
	3. Smell	0.54	0.71	0.76	0.73	0.66	0.6	0.44	0.23	0.15	0.28	0.21	0.24	0.19	0.13
	4. Taste	0.52	0.64	0.76	0.76	0.7	0.71	0.51	0.21	0.23	0.3	0.28	0.25	0.28	0.21
	5. Touch	0.55	0.65	0.74	0.77	0.67	0.79	0.6	0.28	0.24	0.24	0.23	0.34	0.38	0.27
	6. Body	0.59	0.66	0.67	0.76	0.8	0.7	0.6	0.24	0.29	0.3	0.29	0.36	0.43	0.35
	7. Emotion	0.45	0.48	0.51	0.55	0.57	0.69	0.55	0.21	0.23	0.35	0.32	0.33	0.32	0.3
Dream imagery	8. Vision	0.27	0.23	0.21	0.17	0.3	0.27	0.26	0.52	0.54	0.25	0.2	0.55	0.57	0.62
	9. Sound	0.2	0.21	0.21	0.21	0.25	0.29	0.31	0.63	0.49	0.43	0.43	0.54	0.53	0.52
	10. Smell	0.19	0.25	0.3	0.33	0.29	0.31	0.37	0.37	0.64	0.61	0.86	0.41	0.37	0.2
	11. Taste	0.22	0.25	0.28	0.34	0.31	0.32	0.36	0.35	0.6	0.91	0.54	0.41	0.39	0.23
	12. Touch	0.2	0.22	0.2	0.22	0.27	0.27	0.31	0.51	0.58	0.52	0.55	0.58	0.74	0.61
	13. Body	0.19	0.23	0.25	0.2	0.31	0.29	0.28	0.56	0.51	0.45	0.47	0.77	0.59	0.72
	14. Emotion	0.15	0.14	0.13	0.08	0.18	0.16	0.19	0.64	0.52	0.32	0.35	0.6	0.65	0.48

Figure 2. Intercorrelations of study variables in the form of a heat map. The results for measurement time 1 (N_1 = 221) are shown below the diagonal, and the results for measurement time 2 (N_2 = 156) are shown above the diagonal. Monodomain correlations are depicted in italics. Monosense–multidomain correlations are depicted in bold. Retest reliabilities are depicted in bold and italics. The darker the shade of green, the higher the correlation. The underlying multisense–multidomain matrix can be found in the Supplementary Materials. All correlations > 0.15 were significant.

3. Results

3.1. Associations Between the Vividness of Voluntary Mental Imagery and Dream Imagery

The associations of vividness between the different sensory modalities within and between voluntary mental imagery and dream imagery are depicted in Figure 2. The main diagonal provides evidence for satisfactory retest reliabilities in all sensory modalities for both voluntary mental imagery (ranging from 0.55 to 0.76) and dream imagery (ranging from 0.48 to 0.61). Monosense–multidomain correlations are depicted in bold, indicating significant positive correlations between the vividness ratings of different sensory modalities in voluntary mental imagery and the same sensory modalities in dream imagery (Hypothesis 1). As indicated by the darker shades of green in the upper left and lower right quadrants in comparison to the upper right and lower left quadrants, correlations within voluntary mental imagery and within dream imagery, $r = 0.61$, were bigger than across the domains, $r = 0.25$, $z = 4.30$, $p < 0.001$ (Hypothesis 2). Correlation comparisons between all different configurations of the multisense–multidomain matrix can be found in Figure 3A.

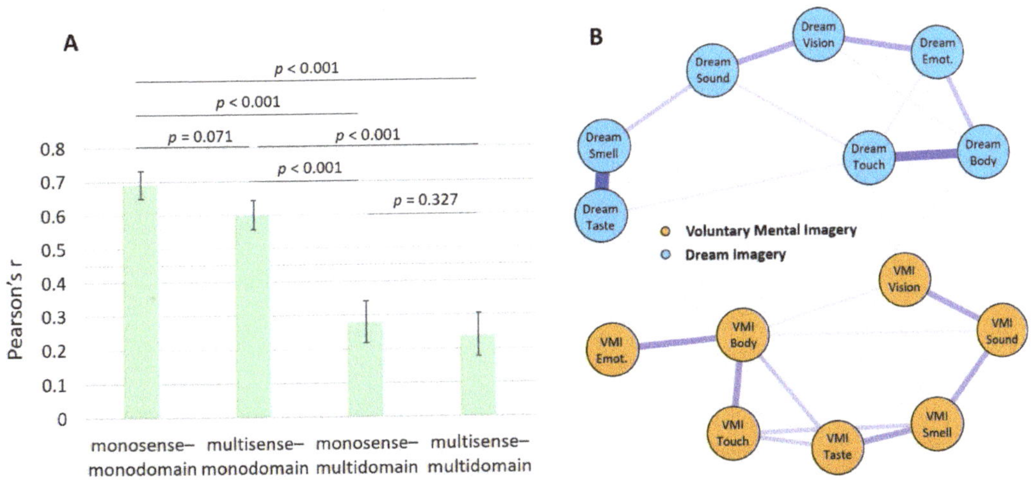

Figure 3. (**A**) Correlation comparisons for the different configurations of the multisense–multidomain matrix. The dotted line indicates the average correlation across all senses and domains. (**B**) Network graph including the subnetworks for the vividness of multisensory voluntary mental imagery (orange) and the vividness of multisensory dream imagery (blue).

Figure 3B depicts the associations between the several modalities of voluntary mental imagery and dream imagery in a network graph using an EBICglasso estimator. In comparison to the sparsity of the total model (= 0.54), the individual models for voluntary mental imagery (*sparsity* = 0.14) and dream imagery (*sparsity* = 0.24) were a lot denser.

Importantly, the association between the vividness of voluntary mental imagery and vividness of dream imagery was moderated by dream recall, as shown by a significantly higher correlation between the vividness of voluntary mental imagery and the vividness of dream imagery when participants recalled their dreams at least once a week, $r = 0.46$, in contrast to less than once in a month, $r = 0.05$, $z = 1.78$, $p = 0.038$ (see Figure 4A) (Hypothesis 5). The same was true for the frequency of lucid dreams: The correlation increased when participants had lucid dreams at least once in a month, $r = 0.62$, in contrast to less than once in a month, $r = 0.28$, $z = 2.62$, $p = 0.004$ (see Figure 4B) (Hypothesis 6). The moderation effect of lucid dreaming could be replicated at measurement time 2, whereas the moderation effect of dream recall was only found descriptively at measurement time 2.

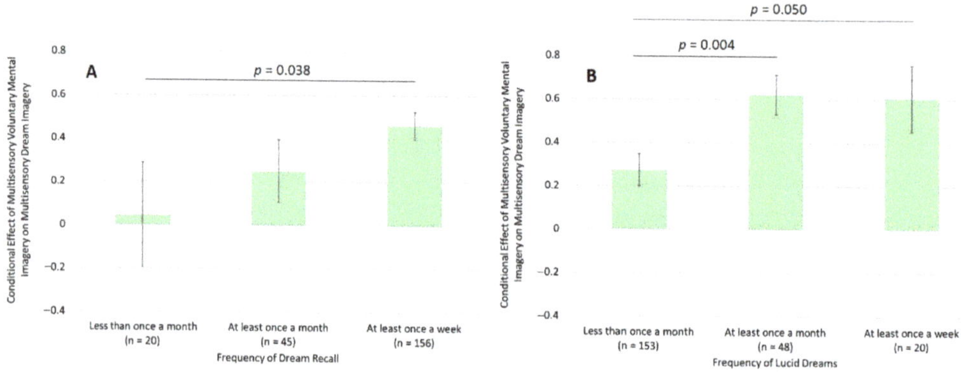

Figure 4. Conditional effects of multisensory voluntary mental imagery vividness on multisensory dream imagery vividness by (**A**) the frequency of dream recall and (**B**) frequency of lucid dreams.

3.2. Mean Differences Between Voluntary Mental Imagery and Dream Imagery

Although multisense–monodomain correlations within voluntary mental imagery, $r = 0.63$, and within dream imagery, $r = 0.55$, did not differ significantly, $z = 1.29$, $p = 0.099$, the mean differences between the domains remain to be investigated. While the vividness of emotional imagery was bigger for dream imagery than for voluntary mental imagery, $t(220) = 8.36$, $p < 0.001$, $d = 0.56$ (Hypothesis 4), the vividness of auditory, $t(220) = 9.39$, $p < 0.001$, $d = 0.63$; olfactory, $t(220) = 17.93$, $p < 0.001$, $d = 1.20$; gustatory, $t(220) = 18.99$, $p < 0.001$, $d = 1.28$; and tactile imagery, $t(220) = 4.91$, $p < 0.001$, $d = 0.33$, were smaller for dream imagery (Hypothesis 3). For visual imagery, $t(220) = 0.20$, $p = 0.984$, $d < 0.01$, and proprioceptive imagery, $t(220) = 0.22$, $p = 0.823$, $d = 0.02$, no significant differences were found. Means and standard errors are depicted in Figure 5. At measurement time 2, aggregated multisense–monodomain correlations within voluntary mental imagery, $r = 0.62$, were significantly bigger than the aggregated multisense–monodomain correlations within dream imagery, $r = 0.46$, $z = 1.99$, $p = 0.023$. This could indicate more pronounced domain general processes in voluntary mental imagery than in dream imagery. Last, the vividness of visual imagery was higher in dream imagery than in voluntary mental imagery, $t(155) = 2.08$, $p = 0.039$, $d = 0.17$.

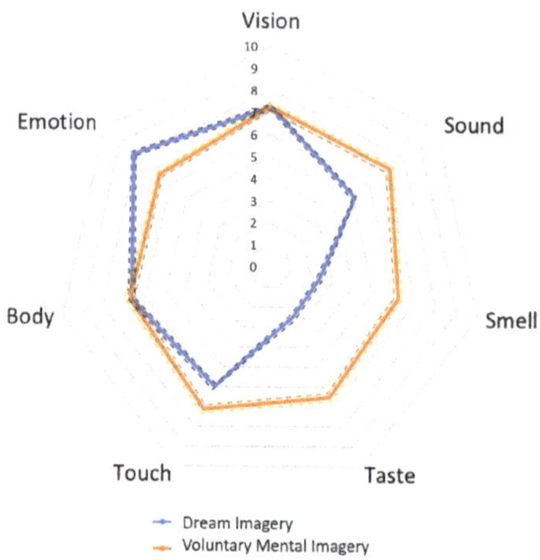

Figure 5. Means (solid lines) and standard errors (dotted lines) for multisensory voluntary mental imagery profiles (orange) and dream imagery profiles (blue).

4. Discussion

Since the vividness of mental imagery is considered a dynamic trait [74], possible state influences on the vividness of voluntary mental imagery and vividness of dream imagery were considered by investigating them at two measurements four weeks apart. Evidence for satisfactory retest reliabilities was found in all sensory modalities for both the vividness of voluntary mental imagery and dream imagery. As most results were replicated at measurement time 2, both measurements are considered together in the following section.

All monosense–multidomain correlations were significant at measurement times 1 and 2 (acceptance of Hypothesis 1). This reflects the neurophysiological overlap between both processes and is consistent with findings that have shown that a loss of visual voluntary mental imagery due to brain damage can also lead to a loss of visual dream experience [12]. It is also consistent with the CESH+, which states that multisensory elements of episodic information must be recombined to produce voluntary and non-voluntary mental imagery [44]. Thus, in addition to the already known neural similarities, our results suggest a phenomenological overlap, supporting a common categorisation of voluntary mental imagery and dream imagery as offline perceptions [9]. The similarities in sensory qualities can also be interpreted as evidence for similar cognitive–perceptual functions across different states of consciousness, which supports the continuity hypothesis of being awake and dreaming [40,81–85].

Next, the monodomain correlations were significantly bigger than the multidomain correlations at measurement times 1 and 2 (acceptance of Hypothesis 2), providing evidence for additional domain-specific processes involved in voluntary mental imagery and dream imagery. In line with this, multisense–multidomain correlations as well as monosense–multidomain correlations were significantly smaller than multisense–monodomain correlations and monosense–monodomain correlations, which can be attributed to the fact that the first correlations were calculated across domains. In contrast, multisense–multidomain correlations were not significantly different from monosense–multidomain correlations and monosense–monodomain correlations were not significantly different from multisense–monodomain correlations, indicating that the correlations were less affected by differences in the sensory modalities. Importantly, the correlations across domains were moderated by dream recall frequency at measurement time 1 (partial acceptance of Hypothesis 5) and by the frequency of lucid dreaming at measurement times 1 and 2 (acceptance of Hypothesis 6). Correlations between the vividness of voluntary mental imagery and vividness of dream imagery were significantly higher when dreams were recalled at least once a week in contrast to less than once a month. This confirms the crucial role of dream recall, since dream recall is a prerequisite to gain access to subjective dream experiences and to investigate the sensory modalities involved in dream imagery. While mental imagery vividness was assessed immediately after participants created the mental image, dream vividness was assessed with a delay, considering the participants' latest dreams.

With regard to lucid dreaming, correlations between the vividness of voluntary imagery and vividness of dream imagery were significantly higher when lucid dreams occurred at least once a month. A higher degree of executive control would be a possible explanation, indicating that a certain degree of executive control and metacognition is present in lucid dreams, but not in non-lucid dreams [86]. The generation of dream imagery in lucid dreaming is therefore more similar to voluntary mental imagery. Reference can be made to [2] which states that dreams are fundamentally subject to the will and, despite the lack of volitional control in dreams, the dreamer can in principle become conscious of his dreaming state. As postulated by [71], lucid dreams can be understood as a hybrid state that once again emphasizes the continuity of consciousness between being awake and dreaming.

Interestingly, the moderation effect of the frequency of lucid dreaming was found at both measurement times, while no significant moderation effect of the frequency of dream recall was found at measurement time 2, although it remained descriptively visible. One explanation could be that the effect of lucid dreaming is more stable, as lucid dreaming involves stronger executive control, which leads to more valid dream reports. Another explanation could be that the frequency of dream recall is more variable than the frequency of lucid dreams due to state influences [61,87]. Unfortunately, we only recorded the frequency of dream recall at measurement time 1, which is why the same value could be invalid at measurement time 2.

Despite the stark associations between the vividness of dream imagery and the vividness of voluntary mental imagery, differences were found in the level of vividness of individual sensory modalities. The vividness of emotional imagery was significantly higher for dream imagery than for voluntary mental imagery at measurement time 1 and 2 (acceptance of Hypothesis 4). This can be explained by the immersive and hallucinatory nature of dreams [60]. While dreaming, emotions are experienced as real, unlike when one tries to imagine a certain emotion and is aware of doing so. The results correspond with previous studies on the affective colouring of dreams, which found that dreams often contain intense emotions, e.g., ref. [58], probably due to consolidation of emotional memory during REM sleep [88].

In contrast to emotional imagery, the vividness of auditory, olfactory, gustatory, and tactile imagery was significantly smaller for dream imagery compared to voluntary mental imagery at both measurement times (acceptance of Hypothesis 3), probably due to higher cognitive control during wakefulness [7]. This is consistent with previous findings showing less frequent reports of gustatory, olfactory, and tactile imagery in dreams, e.g., ref. [36]. On the other hand, at least at measurement time 2, the vividness of visual imagery was higher in dream imagery than in voluntary mental imagery. This is in line with previous studies showing that dreams primarily contain visual imagery [35]. One explanation for the dominance of visual imagery during dreaming could be the continuity hypothesis. Since the visual system is the most important sensory system while being awake, it follows that it is also the most important sensory modality during dreaming (in combination with the immersive quality of dreams [1]). In the same vein, auditory, olfactory, gustatory, and tactile sensations are less pronounced when awake *and* dreaming, e.g., refs. [28,36]. This view is also supported by neuropsychological accounts of the continuity hypothesis that indicate similarities in cognitive and sensory qualities across different states of consciousness [41]. Thus, these results can be regarded as an argument in favour of a continuity between voluntary mental imagery and dream imagery as well.

Finally, the network graphs were denser when voluntary mental imagery and dream imagery were considered separately, indicating that multiple sensory impressions within one state of consciousness are phenomenologically more similar than across both states of consciousness, confirming the other results.

4.1. Implications

Our results suggest that the concept of offline perceptions can be justified on a phenomenological level due to similar multisensory experiences during voluntary mental imagery and dream imagery. We found correlations in terms of vividness in addition to previous results regarding the frequency of specific sensory modalities or shared neural activity. Consequently, certain insights gained from dream imagery or voluntary mental imagery should be transferable from one domain to the other, at least theoretically. Both forms of imagery can be considered as perception-like experiences in the absence of external stimuli. According to this definition, additional mental phenomena can be grouped in this category, such as daydreams and hallucinations [89]. These phenomena should be subject to further investigation, examining neural and phenomenological similarities. Ref. [7] associated hallucinations with dreams and imagination, while [54] suggested that daydreams are particularly similar to voluntary mental imagery. In future studies, a multisensory approach should continue to be used to investigate the relationships of individual sensory modalities in other forms of offline perceptions.

Importantly, the overlap of imagination, dreaming, and perception is of practical relevance to the clinical field. Mental imagery is important in several mental disorders such as PTSD, depression, and nightmare disorders [83,90]. Imagination techniques can reduce dysfunctional mental imagery [91], and nightmares can be treated through approaches such as *Imagery Rehearsal Therapy* [82,92]. These methods are supported by our finding of a strong phenomenological overlap between voluntary mental imagery and dream imagery. Moreover, in line with our results, lucid dreams are proposed as the next step in positively influencing nightmares [93], as they allow for greater cognitive control and are even more similar to dream imagery than voluntary mental imagery while being awake.

4.2. Strengths and Weaknesses

One methodological strength of the present study is the use of the PSI-Q [28], which made it possible to go beyond the assessment of visual imagery, on which research was strongly focused in the past, neglecting other sensory modalities. Analogous to the PSI-Q, a dream questionnaire was developed for the present study. Since the content of dreams cannot be predicted, the dream version of the PSI-Q contained only one general item per modality instead of several items referring to specific dream content. While assessments via single items can be criticized, the questionnaire proved to be retest reliable across both measurement times, further validating our results. However, it should be criticized that the MADRE items concerning dream recall and lucid dreaming were only assessed at the first measurement time. Hence, there was no information about the frequency of dream recall and lucid dreams at the second measurement time. It should also be mentioned that the vividness of voluntary mental imagery and dream imagery was assessed via self-reports, which are subject to response biases. Participants could have inferred their vividness of dream imagery from their vividness of voluntary mental imagery and vice versa or could have generalized from one sensory modality to another. However, despite the possibility of such generalization effects, differences in the vividness levels and correlations were found, rendering our design sufficient for uncovering the expected results. Last, we inferred a link between lucid dreaming and executive control to explain the moderation effect of lucid dreaming. Future research should aim to measure executive functions directly to strengthen our interpretation that executive control might explain the phenomenological overlap between voluntary mental imagery and imagery in lucid dreams.

5. Conclusions

Overall, two separate imagery networks were found, which became increasingly similar after taking dream recall and lucid dreaming into account. On a phenomenological level, this favours a common classification of voluntary mental imagery and dreams as stimulus-independent offline perceptions, using sensory similarity to perception as a criterion. However, even though both phenomena can be categorised as offline perceptions, they are still distinguishable due to other features, such as immersion and executive control.

Supplementary Materials: The following supporting information can be downloaded at https://www.mdpi.com/article/10.3390/vision9020037/s1, Table S1. Intercorrelations of study variables in form of a multisense–multidomain matrix disaggregated by measurement time.

Author Contributions: Conceptualization, M.B. and M.M.; methodology, M.B. and M.M.; software, M.B.; validation, M.M.; formal analysis, M.B. and M.M.; investigation, M.B.; resources, M.M.; data curation, M.B. and M.M.; writing—original draft preparation, M.B. and M.M.; writing—review and editing, M.B. and M.M.; visualization, M.B. and M.M.; supervision, M.M.; project administration, M.M. All authors have read and agreed to the published version of the manuscript.

Funding: This research received no external funding.

Institutional Review Board Statement: As we followed standard procedures in applied psychological research (e.g., informed consent and adherence to the Data Protection Guidelines of the European Union) and did not touch on sensitive topics nor used body invasive procedures or drugs, the procedure and the materials of the study needed no special approval by the Ethics Committee of the Institute for Psychology of the University of Bonn. These guidelines as well as the statement have been approved by the Ethics Committee of the Institute for Psychology of the University of Bonn.

Informed Consent Statement: Digital informed consent was obtained from all participants on the day of participation. This included the voluntary nature of participation, the option of cancelling participation at any time without repercussions, and the use of the anonymized data for scientific purposes.

Data Availability Statement: The data that support the findings of this study are openly available at https://osf.io/ubkv6/?view_only=33331f272a7f4304a7e82cd62393c4fe (accessed on 13 April 2025).

Conflicts of Interest: The authors declare no conflicts of interest.

Abbreviations

The following abbreviations are used in this manuscript:

CESH	Constructed Episodic Simulation Hypothesis
VVIQ	Vividness of Visual Imagery Questionnaire
PSI-Q	Plymouth Sensory Imagery Questionnaire
MADRE	Mannheim Dream Questionnaire

References

1. Windt, J.M. Dreaming Beyond Imagination and Perception. In *The Cambridge Handbook of the Imagination*; Abraham, A., Ed.; Cambridge University Press: Cambridge, UK, 2020; pp. 659–675, ISBN 978-1-108-42924-5.
2. Ichikawa, J.J. Dreaming and Imagination. *Mind Lang.* **2009**, *24*, 103–121. [CrossRef]
3. Ichikawa, J.J. Imagination, Dreaming and Hallucination. In *The Routledge Handbook of Philosophy of Imagination*; Kind, A., Ed.; Routledge: New York, NY, USA, 2016; pp. 149–162.
4. Marks, D.F. Individual Differences in the Vividness of Visual Imagery and Their Effect in Funtion. In *The Function and Nature of Imagery*; Sheehan, P.W., Ed.; Academic Press: New York, NY, USA, 1972; pp. 83–108.
5. Pearson, J. The Human Imagination: The Cognitive Neuroscience of Visual Mental Imagery. *Nat. Rev. Neurosci.* **2019**, *20*, 624–634. [CrossRef] [PubMed]
6. McNorgan, C. A Meta-Analytic Review of Multisensory Imagery Identifies the Neural Correlates of Modality-Specific and Modality-General Imagery. *Front. Hum. Neurosci.* **2012**, *6*, 285. [CrossRef] [PubMed]
7. Waters, F.; Barnby, J.M.; Blom, J.D. Hallucination, Imagery, Dreaming: Reassembling Stimulus-Independent Perceptions Based on Edmund Parish's Classic Misperception Framework: Hallucination, Imagery, Dreaming. *Philos. Trans. R. Soc. B Biol. Sci.* **2021**, *376*, 20190701. [CrossRef]
8. Pearson, J.; Westbrook, F. Phantom Perception: Voluntary and Involuntary Nonretinal Vision. *Trends Cogn. Sci.* **2015**, *19*, 278–284. [CrossRef]
9. Fazekas, P.; Nanay, B.; Pearson, J. Offline Perception: An Introduction. *Philos. Trans. R. Soc. B Biol. Sci.* **2021**, *376*, 20190686. [CrossRef]
10. Domhoff, G.W. The Neural Substrate for Dreaming: Is It a Subsystem of the Default Network? *Conscious Cogn.* **2011**, *20*, 1163–1174. [CrossRef]
11. Eichenlaub, J.-B.; Nicolas, A.; Daltrozzo, J.; Redouté, J.; Costes, N.; Ruby, P. Resting Brain Activity Varies with Dream Recall Frequency Between Subjects. *Neuropsychopharmacology* **2014**, *39*, 1594–1602. [CrossRef]
12. Ffytche, D.H.; Blom, J.D.; Catani, M. Disorders of Visual Perception. *J. Neurol. Neurosurg. Psychiatry* **2010**, *81*, 1280–1287. [CrossRef]

13. Ishai, A.; Sagi, D. Common Mechanisms of Visual Imagery and Perception. *Science* **1995**, *268*, 1772–1774. [CrossRef]
14. Pearson, J.; Clifford, C.W.G.; Tong, F. The Functional Impact of Mental Imagery on Conscious Perception. *Curr. Biol.* **2008**, *18*, 982–986. [CrossRef] [PubMed]
15. Chang, S.; Lewis, D.E.; Pearson, J. The Functional Effects of Color Perception and Color Imagery. *J. Vis.* **2013**, *13*, 4. [CrossRef] [PubMed]
16. Monzel, M.; Keidel, K.; Reuter, M. Imagine, and You Will Find—Lack of Attentional Guidance Through Visual Imagery in Aphantasics. *Atten. Percept. Psychophys.* **2021**, *83*, 2486–2497. [CrossRef] [PubMed]
17. Monzel, M.; Reuter, M. Where's Wanda? The Influence of Visual Imagery Vividness on Visual Search Speed Measured by Means of Hidden Object Pictures. *Atten. Percept. Psychophys.* **2023**, *86*, 22–27. [CrossRef]
18. Laeng, B.; Sulutvedt, U. The Eye Pupil Adjusts to Imaginary Light. *Psychol. Sci.* **2014**, *25*, 188–197. [CrossRef]
19. Horikawa, T.; Kamitani, Y. Generic Decoding of Seen and Imagined Objects Using Hierarchical Visual Features. *Nat. Commun.* **2017**, *8*, 15037. [CrossRef]
20. Tartaglia, E.M.; Bamert, L.; Mast, F.W.; Herzog, M.H. Human Perceptual Learning by Mental Imagery. *Curr. Biol.* **2009**, *19*, 2081–2085. [CrossRef]
21. Knauff, M.; May, E. Mental Imagery, Reasoning, and Blindness. *Q. J. Exp. Psychol.* **2006**, *59*, 161–177. [CrossRef]
22. Dijkstra, N.; Bosch, S.E.; van Gerven, M.A.J. Vividness of Visual Imagery Depends on the Neural Overlap with Perception in Visual Areas. *J. Neurosci.* **2017**, *37*, 1367–1373. [CrossRef]
23. Horikawa, T.; Tamaki, M.; Miyawaki, Y.; Kamitani, Y. Neural Decoding of Visual Imagery during Sleep. *Science* **2013**, *340*, 639–642. [CrossRef]
24. Siclari, F.; Baird, B.; Perogamvros, L.; Bernardi, G.; LaRocque, J.J.; Riedner, B.; Boly, M.; Postle, B.R.; Tononi, G. The Neural Correlates of Dreaming. *Nat. Neurosci.* **2017**, *20*, 872–878. [CrossRef] [PubMed]
25. Leclair-Visonneau, L.; Oudiette, D.; Gaymard, B.; Leu-Semenescu, S.; Arnulf, I. Do the Eyes Scan Dream Images during Rapid Eye Movement Sleep? Evidence from the Rapid Eye Movement Sleep Behaviour Disorder Model. *Brain* **2010**, *133*, 1737–1746. [CrossRef] [PubMed]
26. Lacey, S.; Lawson, R. (Eds.) Introduction. In *Multisensory Imagery*; Springer: New York, NY, USA, 2013; pp. 1–8, ISBN 978-1-4614-5878-4.
27. Zeman, A. Blind Mind's Eye. *Am. Sci. Mag.* **2021**, *109*, 110–117. [CrossRef]
28. Andrade, J.; May, J.; Deeprose, C.; Baugh, S.J.; Ganis, G. Assessing Vividness of Mental Imagery: The Plymouth Sensory Imagery Questionnaire. *Br. J. Psychol.* **2014**, *105*, 547–563. [CrossRef]
29. Sulfaro, A.A.; Robinson, A.K.; Carlson, T.A. Properties of Imagined Experience across Visual, Auditory, and Other Sensory Modalities. *Conscious Cogn.* **2024**, *117*, 103598. [CrossRef]
30. Cui, X.; Jeter, C.B.; Yang, D.; Montague, P.R.; Eagleman, D.M. Vividness of Mental Imagery: Individual Variability Can Be Measured Objectively. *Vis. Res.* **2007**, *47*, 474–478. [CrossRef]
31. Zatorre, R.J.; Halpern, A.R.; Bouffard, M. Mental Reversal of Imagined Melodies: A Role for the Posterior Parietal Cortex. *J. Cogn. Neurosci.* **2010**, *22*, 775–789. [CrossRef]
32. Daselaar, S.M.; Porat, Y.; Huijbers, W.; Pennartz, C.M.A. Modality-Specific and Modality-Independent Components of the Human Imagery System. *Neuroimage* **2010**, *52*, 677–685. [CrossRef]
33. Schwartz, S. A Historical Loop of One Hundred Years: Similarities Between 19th Century and Contemporary Dream Research. *Dreaming* **2000**, *10*, 55–66. [CrossRef]
34. McCarley, R.W.; Hoffmann, E. REM Sleep Dreams and the Activation-Synthesis Hypothesis. *Am. J. Psychiatry* **1981**, *138*, 904–912. [CrossRef]
35. Schredl, M. Characteristics and Contents of Dreams. *Int. Rev. Neurobiol.* **2010**, *92*, 135–154. [PubMed]
36. Okada, H.; Matsuoka, K.; Hatakeyama, T. Individual Differences in the Range of Sensory Modalities Experienced in Dreams. *Dreaming* **2005**, *15*, 106–115. [CrossRef]
37. Domhoff, G.W. The Invasion of the Concept Snatchers: The Origins, Distortions, and Future of the Continuity Hypothesis. *Dreaming* **2017**, *27*, 14–39. [CrossRef]
38. Schredl, M.; Hofmann, F. Continuity between Waking Activities and Dream Activities. *Conscious. Cogn.* **2003**, *12*, 298–308. [CrossRef]
39. Schredl, M.; Reinhard, I. The Continuity between Waking Mood and Dream Emotions: Direct and Second-Order Effects. *Imagin. Cogn. Pers.* **2010**, *29*, 271–282. [CrossRef]
40. Kahan, T.L.; LaBerge, S.P. Dreaming and Waking: Similarities and Differences Revisited. *Conscious. Cogn.* **2011**, *20*, 494–514. [CrossRef]
41. Scarpelli, S.; Alfonsi, V.; Gorgoni, M.; De Gennaro, L. What about Dreams? State of the Art and Open Questions. *J. Sleep Res.* **2022**, *31*, e13609. [CrossRef]

42. Schacter, D.L.; Addis, D.R. Memory and Imagination: Perspectives on Constructive Episodic Simulation. In *The Cambridge Handbook of the Imagination*; Abraham, A., Ed.; Cambridge University Press: Cambridge, UK, 2020; pp. 111–131, ISBN 978-1-108-42924-5.
43. Schacter, D.L.; Addis, D.R. The Cognitive Neuroscience of Constructive Memory: Remembering the Past and Imagining the Future. *Philos. Trans. R. Soc. B Biol. Sci.* **2007**, *362*, 773–786. [CrossRef]
44. Blomkvist, A. Aphantasia: In Search of a Theory. *Mind Lang.* **2023**, *38*, 866–888. [CrossRef]
45. Danker, J.F.; Anderson, J.R. The Ghosts of Brain States Past: Remembering Reactivates the Brain Regions Engaged during Encoding. *Psychol. Bull.* **2010**, *136*, 87–102. [CrossRef]
46. Spanò, G.; Pizzamiglio, G.; McCormick, C.; Clark, I.A.; De Felice, S.; Miller, T.D.; Edgin, J.O.; Rosenthal, C.R.; Maguire, E.A. Dreaming with Hippocampal Damage. *eLife* **2020**, *9*, e56211. [CrossRef] [PubMed]
47. Dawes, A.J.; Keogh, R.; Andrillon, T.; Pearson, J. A Cognitive Profile of Multi-Sensory Imagery, Memory and Dreaming in Aphantasia. *Sci. Rep.* **2020**, *10*, 10022. [CrossRef] [PubMed]
48. Zeman, A.; Milton, F.; Della Sala, S.; Dewar, M.; Frayling, T.; Gaddum, J.; Hattersley, A.; Heuerman-Williamson, B.; Jones, K.; MacKisack, M.; et al. Phantasia–The Psychological Significance of Lifelong Visual Imagery Vividness Extremes. *Cortex* **2020**, *130*, 426–440. [CrossRef] [PubMed]
49. Monzel, M.; Leelaarporn, P.; Lutz, T.; Schultz, J.; Brunheim, S.; Reuter, M.; McCormick, C. Hippocampal-Occipital Connectivity Reflects Autobiographical Memory Deficits in Aphantasia. *eLife* **2023**, *13*, RP94916. [CrossRef]
50. Jungmann, S.M.; Becker, F.; Witthöft, M. Erfassung Der Lebendigkeit Mentaler Vorstellungsbilder. *Diagnostica* **2022**, *68*, 125–136. [CrossRef]
51. Okada, H.; Matsuoka, K.; Hatakeyama, T. Dream-Recall Frequency and Waking Imagery. *Percept. Mot. Ski.* **2000**, *91*, 759–766. [CrossRef]
52. Revonsuo, A.; Tuominen, J.; Valli, K. Avatars in the Machine: Dreaming as a Simulation of Social Reality. In *Open MIND*; MIND Group: Frankfurt am Main, Germany, 2016; Volume 2, pp. 1295–1322. [CrossRef]
53. Domhoff, G.W. Dreaming Is Imagination Roaming Freely, Based on Embodied Simulation and Subserved by an Unconstrained Default Network. In *The Cambridge Handbook of the Imagination*; Abraham, A., Ed.; Cambridge University Press: Cambridge, UK, 2020; pp. 676–691, ISBN 978-1-108-42924-5.
54. Occhionero, M.; Cicogna, P. Phenomenal Consciousness in Dreams and in Mind Wandering. *Philos. Psychol.* **2016**, *29*, 958–966. [CrossRef]
55. McCormick, C.; Ciaramelli, E.; De Luca, F.; Maguire, E.A. Comparing and Contrasting the Cognitive Effects of Hippocampal and Ventromedial Prefrontal Cortex Damage: A Review of Human Lesion Studies. *Neuroscience* **2018**, *374*, 295–318. [CrossRef]
56. Yomogida, Y. Mental Visual Synthesis Is Originated in the Fronto-Temporal Network of the Left Hemisphere. *Cereb. Cortex* **2004**, *14*, 1376–1383. [CrossRef]
57. Buda, M.; Fornito, A.; Bergström, Z.M.; Simons, J.S. A Specific Brain Structural Basis for Individual Differences in Reality Monitoring. *J. Neurosci.* **2011**, *31*, 14308–14313. [CrossRef]
58. Fosse, R.; Stickgold, R.; Hobson, J.A. The Mind in REM Sleep: Reports of Emotional Experience. *Sleep* **2001**, *24*, 947–955. [CrossRef] [PubMed]
59. Blackwell, S.E. Emotional Mental Imagery. In *The Cambridge Handbook of the Imagination*; Abraham, A., Ed.; Cambridge University Press: New York, NY, USA, 2020; pp. 241–257.
60. Windt, J.M. The Immersive Spatiotemporal Hallucination Model of Dreaming. *Phenomenol. Cogn. Sci.* **2010**, *9*, 295–316. [CrossRef]
61. Schredl, M. Traumerinnerung. In *Traum*; utb GmbH: Stuttgart, Germany, 2008; pp. 13–21, ISBN 9783825230050.
62. Schredl, M. Dream Recall: Models Ans Empirical Data. In *The New Science of Dreaming: Vol. 2 Content, Recall, and Personality Correlates*; Barrett, D., McNamara, P., Eds.; Praeger Publishers; Greenwood Publishing Group: Westport, CT, USA, 2007; pp. 79–114.
63. Fazekas, P.; Nemeth, G.; Overgaard, M. White Dreams Are Made of Colours: What Studying Contentless Dreams Can Teach about the Neural Basis of Dreaming and Conscious Experiences. *Sleep Med. Rev.* **2019**, *43*, 84–91. [CrossRef] [PubMed]
64. Hiscock, M.; Cohen, D.B. Visual Imagery and Dream Recall. *J. Res. Pers.* **1973**, *7*, 179–188. [CrossRef]
65. Filevich, E.; Dresler, M.; Brick, T.R.; Kühn, S. Metacognitive Mechanisms Underlying Lucid Dreaming. *J. Neurosci.* **2015**, *35*, 1082–1088. [CrossRef]
66. Dresler, M.; Wehrle, R.; Spoormaker, V.I.; Koch, S.P.; Holsboer, F.; Steiger, A.; Obrig, H.; Sämann, P.G.; Czisch, M. Neural Correlates of Dream Lucidity Obtained from Contrasting Lucid versus Non-Lucid REM Sleep: A Combined EEG/FMRI Case Study. *Sleep* **2012**, *35*, 1017–1020. [CrossRef]
67. Cavanna, A.E.; Trimble, M.R. The Precuneus: A Review of Its Functional Anatomy and Behavioural Correlates. *Brain* **2006**, *129*, 564–583. [CrossRef]
68. Christoff, K.; Ream, J.M.; Geddes, L.P.T.; Gabrieli, J.D.E. Evaluating Self-Generated Information: Anterior Prefrontal Contributions to Human Cognition. *Behav. Neurosci.* **2003**, *117*, 1161–1168. [CrossRef]

69. Schmitz, T.W.; Kawahara-Baccus, T.N.; Johnson, S.C. Metacognitive Evaluation, Self-Relevance, and the Right Prefrontal Cortex. *Neuroimage* **2004**, *22*, 941–947. [CrossRef]
70. Maquet, P.; Péters, J.-M.; Aerts, J.; Delfiore, G.; Degueldre, C.; Luxen, A.; Franck, G. Functional Neuroanatomy of Human Rapid-Eye-Movement Sleep and Dreaming. *Nature* **1996**, *383*, 163–166. [CrossRef]
71. Voss, U.; Holzmann, R.; Tuin, I.; Hobson, J.A. Lucid Dreaming: A State of Consciousness with Features of Both Waking and Non-Lucid Dreaming. *Sleep* **2009**, *32*, 1191–1200. [CrossRef] [PubMed]
72. Denis, D.; Poerio, G.L. Terror and Bliss? Commonalities and Distinctions between Sleep Paralysis, Lucid Dreaming, and Their Associations with Waking Life Experiences. *J. Sleep Res.* **2017**, *26*, 38–47. [CrossRef] [PubMed]
73. Gackenbach, J.; Schillig, B. Lucid Dreams: The Content of Consious Awareness of Dreaming during the Dream. *J. Ment. Imag.* **1983**, *7*, 1–13.
74. McKelvie, S.J. The VVIQ as a Psychometric Test of Individual Differences in Visual Imagery Vividness: A Critical Quantitative Review and Plea for Direction. *J. Ment. Imag.* **1995**, *19*, 1–106.
75. Callow, N.; Roberts, R.; Fawkes, J.Z. Effects of Dynamic and Static Imagery on Vividness of Imagery, Skiing Performance, and Confidence. *J. Imag. Res. Sport Phys. Act.* **2006**, *1*, 2006. [CrossRef]
76. Schredl, M. Factors Affecting the Continuity between Waking and Dreaming: Emotional Intensity and Emotional Tone of the Waking-Life Event. *Sleep Hypn.* **2006**, *8*, 1–5.
77. Schredl, M.; Berres, S.; Klingauf, A.; Schellhaas, S.; Göritz, A.S. The Mannheim Dream Questionnaire (MADRE): Retest Reliability, Age and Gender Effects. *Int. J. Dream Res.* **2014**, *7*, 141–147.
78. Leiner, D.J. SoSci Survey (Version 3.2.50) [Computer Software]. 2019. Available online: https://www.soscisurvey.de (accessed on 20 March 2025).
79. IBM Corp. IBM SPSS Statistics for Windows (Version 27.0) [Computer software]. 2022. Available online: https://www.ibm.com/products/spss-statistics (accessed on 20 March 2025).
80. Lenhard, W.; Lenhard, A. Psychometrica: Signifkanztests bei Korrelationen. 2014. Available online: http://www.psychometrica.de/correlation.html (accessed on 20 March 2025).
81. JASP Team. JASP (Version 0.16.4) [Computer Software]. 2022. Available online: https://jasp-stats.org/ (accessed on 20 March 2025).
82. Krakow, B.; Kellner, R.; Pathak, D.; Lambert, L. Imagery Rehearsal Treatment for Chronic Nightmares. *Behav. Res. Ther.* **1995**, *33*, 837–843. [CrossRef]
83. Holmes, E.A.; Lang, T.J.; Moulds, M.L.; Steele, A.M. Prospective and Positive Mental Imagery Deficits in Dysphoria. *Behav. Res. Ther.* **2008**, *46*, 976–981. [CrossRef]
84. Cohen, D.B.; MacNeilage, P.F. A Test of the Salience Hypothesis of Dream Recall. *J. Consult. Clin. Psychol.* **1974**, *42*, 699–703. [CrossRef]
85. Keogh, R.; Pearson, J.; Zeman, A. Aphantasia: The science of visual imagery extremes. *Handb. Clin. Neurol.* **2021**, *178*, 277–296. [PubMed]
86. La Berge, S.P. Lucid Dreaming as a Learnable Skill: A Case Study. *Percept. Mot. Skills* **1980**, *51*, 1039–1042. [CrossRef]
87. Schredl, M. Luzide Träume. In *Traum*; utb GmbH: Stuttgart, Germany, 2008; pp. 73–80, ISBN 9783825230050.
88. Malinowski, J.E.; Horton, C.L. Dreams Reflect Nocturnal Cognitive Processes: Early-Night Dreams Are More Continuous with Waking Life, and Late-Night Dreams Are More Emotional and Hyperassociative. *Conscious. Cogn.* **2021**, *88*, 103071. [CrossRef] [PubMed]
89. Nanay, B. Unconscious Mental Imagery. *Philos. Trans. R. Soc. B Biol. Sci.* **2021**, *376*, 20190689. [CrossRef]
90. Pearson, J.; Naselaris, T.; Holmes, E.A.; Kosslyn, S.M. Mental Imagery: Functional Mechanisms and Clinical Applications. *Trends Cogn. Sci.* **2015**, *19*, 590–602. [CrossRef]
91. Williams, A.D.; Blackwell, S.E.; Mackenzie, A.; Holmes, E.A.; Andrews, G. Combining Imagination and Reason in the Treatment of Depression: A Randomized Controlled Trial of Internet-Based Cognitive-Bias Modification and Internet-CBT for Depression. *J. Consult. Clin. Psychol.* **2013**, *81*, 793–799. [CrossRef]
92. Schredl, M.; Hoppe, J. Etiology and Treatment of Nightmares. *Somnologie* **2019**, *23*, 49–57. [CrossRef]
93. Gott, J.; Bovy, L.; Peters, E.; Tzioridou, S.; Meo, S.; Demirel, C.; Esfahani, M.J.; Oliveira, P.R.; Houweling, T.; Orticoni, A.; et al. Virtual Reality Training of Lucid Dreaming: VR Training of Lucid Dreaming. *Philos. Trans. R. Soc. B Biol. Sci.* **2021**, *376*, 20190697. [CrossRef]

vision

MDPI

Article

Stronger Short-Term Memory, Larger Hippocampi and Area V1 in People with High VVIQ Scores

David F. Marks

Arles, Bouches-du-Rhône, Provence-Alpes-Côte d'Azur, 13200, France; dfmarksphd@gmail.com

Abstract

Reports of individual differences in vividness of visual mental imagery (VMI) scores raise complex questions: Are Vividness of Visual Imagery Questionnaire (VVIQ) score differences actually measuring anything? What functions do these differences serve? What is their neurological foundation? A new analysis examined visual short-term memory (VSTM) and volumes of the hippocampi, primary visual cortices, and other cortical regions among vivid and non-vivid visual imagers. In a sample of 53 volunteers aged 54 to 80 with MRI scans, the performance of ten Low VVIQ scorers was compared to that of ten High VVIQ scorers. The groups included an aphantasic with a minimum VVIQ score and a hyperphantasic with a maximum VVIQ score. The study examined volumes for 12 hippocampal subfields, 11 fields implicated in visual mental imagery including area V1 and the fusiform gyrus, and 7 motor regions. In comparison to the Low VVIQ group, High VVIQ group yielded: (i) significantly more accurate VSTM performance; and (ii) significantly larger volumes of the hippocampi and primary visual cortex. Across 47 brain regions, the average volume for the High VVIQ group exceeded that of the Low VVIQ group by 11 percent. For 47 subfields, the volumes of the hphantasic exceeded those of the aphantasic person by an average of 57 percent. Females had more accurate visual short-term memory than males and younger people were more accurate than older people. The larger visual memory capacity of females was unmatched by larger regional volume differences, which suggests that the sex difference in visual memory is caused by factors other than cortical regional size. The study confirms the existence of robust empirical associations between VMI vividness, short-term memory, regional volume of hippocampal subfields and area V1.

Keywords: VVIQ; visual mental imagery; short-term memory; hippocampus; cortex; individual differences; aphantasia; hyperphantasia

1. Introduction

In this study we are interested in the role of visual and other mental imagery in cognition, imagination, and action, directed towards the present or anticipated states of the external world. Here our focus is individual differences. This paper reports an investigation into the voluntary form of visual mental imagery (VMI) that occurs in the majority of healthy people throughout their waking lives. VMI refers to a capability within subjective experience to "see" an object, scene, or activity in the absence of the stimuli normally present when that object, scene, or activity is perceived. The well-documented, individual differences in waking VMI capacity indexed by vividness raise fundamental questions about mind–brain connections [1]. VMI vividness differences may be distinctive, but are they actually measuring anything, what functions do they serve, and what is their neurological foundation? The aim of the present study is to provide a clear answer to these three questions via an investigation of the relationship between VMI vividness scores and (1) short-term visual memory (VSTM) and (2) volumes of brain areas understood to mediate VMI.

A large body of research is consistent with the view that imagery is functionally equivalent to, but not identical to, perception showing similar anatomical patterning of neural activity in the cerebral cortex and subcortical regions, especially the hippocampus. As in perception, a preparedness for action with feelings and motives are key components of mental imagery experience [1]. Over the past 50 years much of the literature on individual differences in VMI has focused on the construct of vividness using a self-report, subjective measure called the *Vividness of Visual Imagery Questionnaire* (VVIQ) [1–25]. The vividness of a visual image is defined as a combination of clarity, colourfulness, and liveliness, where clarity is defined by brightness and sharpness, colourfulness by image saturation, and liveliness by vivacity, animation, feeling, solidity, projection and metamorphosis [2,3].

Mental imagery in any modality conveys intentionality—mental images are laden with vigour and feeling such as anger, anticipation, appreciation, calm, disgust, fear, or tranquillity. Not overlooking the vivid imagery of the sleep-state in dreaming, another study within this Special Issue suggests that dreams contain emotional and visual imagery vividness at significantly higher levels than is evident in waking mental imagery (Bilzer & Monzel, 2025) [4]. In everyday life, the enactive and affective attributes of mental imagery are capable of moving readers of literature to experience fictional events and characters as if they are "real". Thus, VMI is closely associated with feeling and aesthetic appreciation of poetry, haiku, or sonnets and enhances one's appreciation of music [5]. Recent theorizing in the General Theory of Behaviour emphasizes the role of mental imagery as a catalyst or container of feeling and motivation [6].

Here, I use an alternative methodology to present a re-analysis of an original correlational dataset from an investigation of the role of VMI in visual short-term memory (VSTM) and its foundation in brain systems, which include the subfields of the hippocampi and several other brain regions. Brain areas thought to participate in episodic memory and VMI are compared to motor control regions that, ordinarily, are not implicated in VMI functions. Previous research found that High VVIQ scorers have superior picture recall compared to Low VVIQ scorers [2,7], with females showing superior picture recall to males. The superior memory of vivid visual imagers has been independently replicated [8–16]. Rodway, Gillies, and Schepman (2006) [17] found high vividness participants more accurate in detecting salient changes to pictures compared to low vividness participants. An aphantasic who lacked the capacity for voluntary VMI showed worse visual working memory than the controls when high precision was required [18].

Some researchers assert that VVIQ differences are an artefact of imagers making false attributions about their capability, e.g., [19–21]. Yet, in spite of this criticism, experiments using the VVIQ [2,7] have repeatedly revealed robust, replicable differences between High and Low VVIQ-scoring subjects across different ages, sexes, and cultures [1,22]. The VVIQ is commonly used to identify "aphantasia" [23] and "hyperphantasia" [24], which is giving renewed impetus to the field [25]. A systematic review in this Special Issue emphasizes the need to "refine the definition and diagnosis of aphantasia" [26].

In one explanation of the distinctive individual differences in mental imagery ability, it has been suggested that the normal distribution of VMI vividness across the population may involve a subclinical form of "disconnection" [27–29]. This view received early support from electroencephalographic (EEG) studies that indicated a left hemisphere focus for VMI in subjects having vivid visual imagery with higher brain activation in the left parieto-occipital cortex, which was found to be lacking in matched, non-vivid imagers [30–32]. These EEG findings were consistent with reports of patients with brain lesions in a subfield of the posterior left hemisphere who appeared to lack the capacity of VMI generation [28,33].

Activation of the primary visual cortex is normally thought to be predictive of imagery vividness in healthy individuals [34–38]. However, different findings between laboratory studies with healthy volunteers and clinical case studies are reported. One clinical study found that the early visual cortex was unnecessary for VMI and, at the same time, found the left fusiform gyrus to be crucial [39]. This difference in opinion about the cortical foundation of VMI remains unresolved, although Dijkstra [40] in this Special Issue has suggested one possible solution. Bergmann, Genç, Kohler et al. (2015) [41] found a negative relationship between primary visual cortex (V1) surface area and sensory imagery strength but positive relationships between V1 and imagery precision. Other varying results have been obtained, including [42,43], who found that left precuneus activation in both static and dynamic imagery tasks was positively correlated with VVIQ scores.

Connectivity is clearly crucial. Tullo et al. (2022) [44] found the occipito-medial region has an inhibitory influence on temporal regions, and an excitatory influence of more anterior on more medial and posterior brain regions. VMI vividness was associated with connection strength from an occipital area to a parahippocampal area, especially in the left hemisphere. Ref. [45] also report increased efficiency and clustering in left inferior temporal regions in individuals with more vivid visual imagery. Thus, area V1, the left IT region, and the anterior and posterior hippocampus all appear crucial to VMI vividness [46].

The study presented here examines differences between people with Low and High VVIQ scores and gender differences in visual STM. The finding of superior visual memory of females [47] has been replicated in several contexts [48–57] while tasks involving visuospatial memory tend to favour males [58]. Relative to brain size, findings are mixed [59–67], suggesting that the sex difference observed on the structural level does not directly relate to sex differences in cognition, which likely depend less on how the brain is structured but on functional organization and connectivity. Dalton, Zeidman, McCormick, and Maguire (2018) [68] found that different parts of the hippocampus, together with distinct cortical regions, are recruited for scene construction or non-scene-evoking associative processing. Connections between different parts of the hippocampus and the visual–perceptual cortex appear particularly significant to the generation of VMI. Superior memory at different ages is strongly related to hippocampal structure and function [69]. Common observation and clinical testing indicate that ageing generally brings memory into decline, yet 80-year-old-plus so-called "super-agers" apparently have an episodic memory capability similar to that of 50- to 65-year-olds [70].

As noted, aphantasia refers to reduced or complete absence of voluntary VMI, while hyperphantasia refers to extremely high VMI availability and vividness [23,24]. Aphantasic subjects are selected for studies in different ways with VVIQ scores anywhere from 16 points (complete absence of visual imagery) to 32 points ("vague and dim" visual imagery, incomplete absence), e.g., [71,72]. Monzel et al. (2024) [72] investigated whether the connectivity of the hippocampus in aphantasia differs from that of controls. They investigated autobiographical memory (AM) of 14 "congenital aphantasics" and 16 controls to see how the hippocampus and visual–perceptual cortices interact during AM re-experiencing. Aphantasics showed almost no functional connectivity between the hippocampus and the visual–perceptual areas. This finding aligns with a top-down hierarchical theory of AM in which the hippocampus initiates retrieval processes in the V1 area to retrieve visual–perceptual details (Blomkvist [73]).

The impetus for the present study was a publication by Tabi et al. (2022) [74], which examined the association between VMI, short-term memory (STM), and hippocampal and primary visual cortical volumes using a correlational approach. To the best of this author's knowledge, the study by Tabi et al. [74] is the only existing study to investigate brain regional volumes among individuals with differing VVIQ scores. In their Study 1, Tabi et al. found no evidence of a relationship between VMI and STM performance. However, significant positive correlations were found between VVIQ scores, hippocampal subfield volumes, and volumes of primary visual cortex. VSTM performance was significantly impaired in patients with AD but the VMI scores of the patients with AD were comparable to those of the age-matched elderly controls and patients with HD. Although hippocampal volumes were reduced in AD patients, there was no reduction in their VVIQ scores, suggesting that vividness may be correlated with hippocampal and visual cortex volume in healthy people but not with VSTM [74]. Here, an alternative perspective on the VVIQ/VSTM association is presented.

For this study, the dataset provided by [74] was re-analyzed using a protocol that places a predetermined number (in this case, 10) of extreme-scoring subjects into High and Low VVIQ scoring groups. Previous studies of VVIQ differences in cognition, e.g., [2,7,16] suggested that the "extreme groups" protocol would be more sensitive in detecting vividness group effects than correlational analyses conducted on an entirely unselected sample.

Another novel feature of the present study is that it is the first to investigate anatomical differences in brain regional sizes of aphantasic (aphant) and hyperphantasic (hphant) individuals.

The study tested 16 clearly stated and falsifiable hypotheses:

H1: *People with vivid visual imagery have greater VSTM than people with non-vivid visual imagery.*

H2: *Females have greater VSTM than males.*

H3: *Younger people have greater VSTM than older people.*

H4: *In brain areas related to VMI functioning (i.e., hippocampi, V1, fusiform gyrus), the High VVIQ group has larger volumes than the Low VVIQ group.*

H5: *In brain areas related to VMI functioning, females have larger volumes than males.*

H6: *In brain areas related to VMI functioning, younger people have larger volumes than older people.*

H7: *In brain areas unrelated to VMI functioning (i.e., amygdala, Brodmann area 4), regional volume does not vary with VVIQ group.*

H8: *In brain areas unrelated to VMI functioning, regional volume does not vary with gender.*

H9: *In brain areas unrelated to VMI functioning, regional volume does not vary with age.*

H10: *An aphant has weaker than average VSTM.*

H11: *An hphant has stronger than average VSTM.*

H12: *An hphant has stronger VSTM than an aphant.*

H13: *An aphant has smaller than average VMI functioning volumes.*

H14: *An hphant has larger than average VMI functioning volumes.*

H15: *An hphant has larger VMI functioning volumes than an aphant.*

H16: *Volume sizes across groups and individuals follow a predictable sequence: hphant score highest, High VVIQ group mean, entire sample mean, Low VVIQ group mean, aphant score lowest.*

2. Materials and Methods

The materials and methods are described in detail elsewhere [74]. (The methods, participants, data collection, and curation used in the preparation of this article were contributed by Tabi et al. (2022) [74] at the University of Oxford and are available at: https://osf.io/q37vn/ accessed on 26 June 2025. Full details are available at: https://www.sciencedirect.com/science/article/pii/S0010945221003488 accessed on 26 June 2025. Younes Adam Tabi kindly gave permission for this re-analysis but did not participate in the analyses or writing of this article.) The following is a condensed description.

2.1. Short-Term Memory Task

A "What was where?" object-location VSTM task (Figure 1) was used to measure VSTM [75]. At a viewing distance of approximately 30 cm, the task was performed on a touch-sensitive screen (iPad version 9.3.5 (13G36), model MGTX2B/A) with a 1536 × 2048-pixel matrix. Stimuli were drawn from a library of foils and presented on a black background, randomly selected without repetition.

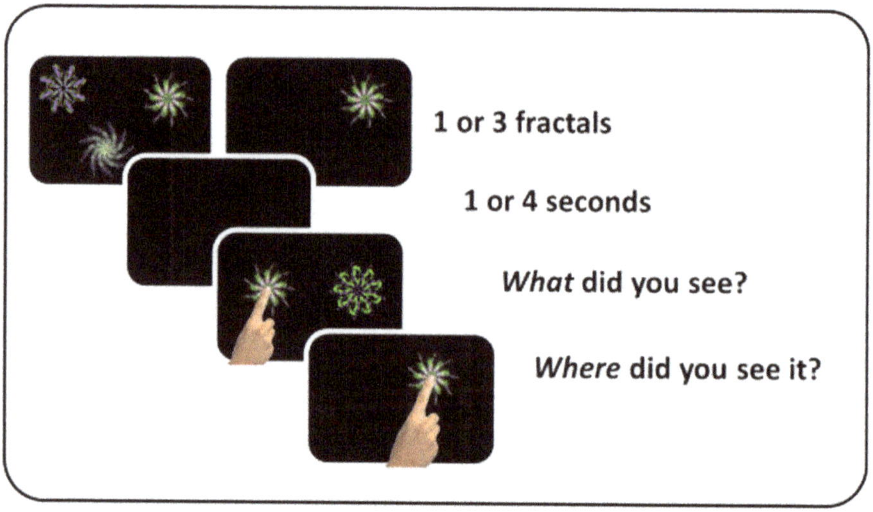

Figure 1. Task schematic of the visual short-term memory task. Participants saw one or three fractal designs for a period of 1 s per fractal. After a blank interval of 1 or 4 s, one of the original fractals reappeared together with a distractor fractal. Participants had to select the fractal which they thought had been previously presented (Identification accuracy) and drag it to where they recalled it had previously appeared (Localisation performance). Reproduced under a Creative Commons licence, originally published by Tabi et al. [74].

Participants were presented with either one or three random shapes in the form of fractals. If one fractal was presented, it appeared for 1 s; if three fractals were presented, they appeared for 3 s. The mean duration for encoding was always 1 s per item. After a delay lasting 1 or 4 s, participants were asked to select the shape they had previously seen from two shapes presented at the centre of the screen, one of which was a never-seen foil. To select the correct shape, they were required to touch it, which provided discrete measures of Identification Accuracy and Response Time. Then they were requested to drag the selected item across the screen to its remembered location, which allowed measurement of Localisation Performance on an analogue scale: the distance between the actual location where the fractal had appeared and the location remembered by the participant.

When the person stopped dragging the chosen shape, a "Done" button appeared at the bottom centre of the screen. If participants started dragging the shape again, this disappeared and then reappeared upon release of the shape. The final position was supported by them pressing the "Done" button. To continue with the next trial, they simply pressed a "Next" button.

After 19 practice trials, the volunteers performed 120 trials over three blocks, 30 trials per condition equally distributed across blocks. The four conditions consisted of either 1 or 3 shapes combined with delays of either 1 or 4 s. The following parameters were available:

- Identification Accuracy: Trials in which the correct shape was identified out of two divided by the total number of trials of each condition. Trials in which participants did not identify the correct item were excluded from further analysis.
- Response Time: Time taken to point to the correct shape.
- Localisation Performance: Distance between the response and the original target's location.
- Misbinding; The probability that a participant can correctly remember the appearances and locations of the shapes at test but confuses or *"misbinds"* these locations and appearances, resulting in the report of another shape's location for a correctly identified shape.
- Guessing, Guessing indicates a random response likely given when the viewer has forgotten any precise information about the target.

For each trial, the distances between the response and (i) the target, (ii) the closest non-target, and (iii) another random trial's non-target were compared with one another. If the distance to the target was closest, this was counted as a target response. If the distance to the non-target was closest, this was counted as a misbinding. If the distance to the random trial's non-target was closest, this was counted as a guessing. For each trial, this procedure was repeated 5000 times with a random trial's non-target each time. Thus, probabilities for each trial for misbinding and guessing could be calculated.

2.2. *Vividness of Visual Imagery Questionnaire*

All participants were tested on the VVIQ [2,7]. The questionnaire consists of 16 items which participants score between 1 and 5 according to how vividly they can imagine a familiar person, a shop, the sky, or a countryside scene, leading to a total score between 16 and 80.

2.3. *MRI Analysis*

A 3T Siemens Magnetom Verio syngo scanner was used to acquire T1-weighted volumetric images through a magnetization prepared rapid gradient echo protocol (MPRAGE) in sagittal orientation (TR = 2000 msec, TE = 1.94 msec, TI = 880 msec, Flip angle = 8°, FOV read = 256 mm, Voxel size = 1.0 × 1.0 × 1.0 mm). The same machine was used to record T2-FLAIR images.

FSL FIRST [76] was used to generate hippocampal and amygdala volume for both hemispheres from the T1 anatomical images. These volumes were corrected for age and total intracranial volume calculated using FSL SIENAX, which estimates total brain tissue volume from a single image, normalized for skull size. It strips non-brain tissue and uses the brain and skull images to estimate the scaling between the subject's image and standard space. It then runs tissue segmentation to estimate the volume of brain tissue, and multiplies this by the estimated scaling factor, to reduce head size-related variability between subjects.

Grey matter volume of BA4 and V1 was calculated through the standard Freesurfer pipeline (cortical reconstruction and volumetric segmentation: http://surfer.nmr.mgh.harvard.edu/ (accessed on 1 July 2025) corrected for age and total intracranial volume. In addition, hippocampal subfields were decomposed using the software's subfield pipeline based on T1 and T2 image inputs.

2.4. Availability of Data

All data are provided at: https://osf.io/q37vn/ (accessed on 1 July 2025).

2.5. Participants

Participants gave their informed consent to be involved in the study, which was approved by the local ethics committee [70]. The sample consisted of 53 healthy, middle- to older-aged participants who completed an MRI scan, the VSTM task, and VVIQ. (Although Tabi et al. [74] state that their sample size was 56, the MRI dataset available at the online repository contains data for only 53 subjects.) In following the extreme groups protocol, it was decided in advance to select and compare subgroups consisting of the ten highest VVIQ and ten lowest VVIQ scorers. A gap of 21 points separated the Low VVIQ (16–52) and the High VVIQ (73–80) groups (Table 1). The High VVIQ group contained six females, one of whom scored 80 (an "hphant" who rated all 16 VVIQ items at level five—"Perfectly clear and as vivid as normal vision"). The Low VVIQ group included five females, one having a minimum score of 16 (an "aphant" who rated all 16 VVIQ items at level one—"No image at all, you only "know" that you are thinking of an object"). Figure 2 shows the probability distributions of the samples. Table 1 describes the five samples of the participants.

Table 1. The sample sizes, gender split, age means and SD, and VVIQ mean scores and SD.

Sample	N	Female: Male	Age Mean (SD)	VVIQ Mean (SD)
Whole sample	53	29:24	68.00 (7.20)	62.23 (12.53)
High VVIQ group	10	6:4	67.40 (7.83)	72.90 (9.46)
Low VVIQ group	10	5:5	68.70 (6.77)	44.50 (15.51)
Younger group (54–62 years)	10	3:7	57.9 (2.60)	58.80 (14.93)
Older group (76–80 years)	11	4:7	78.0 (1.70)	65.45 (10.55)

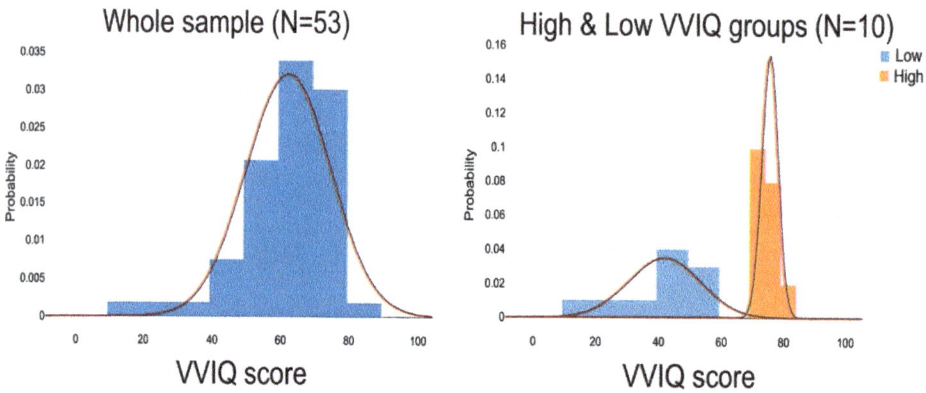

Figure 2. Probability distributions of the sample (left side) and the High VVIQ and Low VVIQ groups.

For analysis of the effect of age, the 11 oldest participants (aged 76 to 80 years) and the 10 youngest participants (aged 54 to 62) were selected, leaving an age gap of 14 years in between (Figure 3). The inclusion of three individuals tied at age 76 required N for the older group to be 11 instead of 10. The mean ages of the younger and older groups were 57.9 (SD = 2.60) and 78.0 (SD = 1.70), respectively (Figure 3; Table 1). The mean age of the original sample was 68.0 (SD = 7.2, MIN = 54, MAX = 80) of whom 29/53 (54.7%) were female.

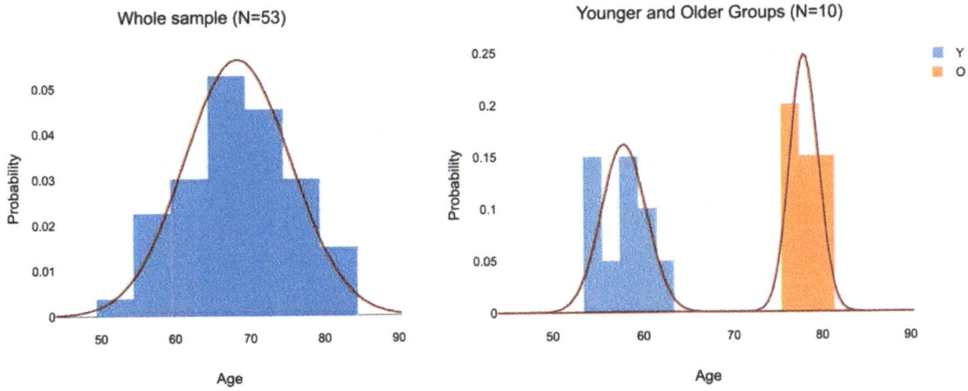

Figure 3. Probability distributions of the whole sample (left side) and the younger and older subgroups.

2.6. Ethical Approval

All participants gave their informed consent to be involved in the study, which was approved by the local ethics committee at the University of Oxford, as reported by [74].

3. Results

The results for all group and individual comparisons are presented in Table S1 (Supplementary file). Normality testing of the dependent variables indicated that transformation of scores was unnecessary.

3.1. VVIQ Scores

Table S1 (Supplementary file) presents the mean age and VVIQ scores with standard deviations and outcomes for the members of the whole sample, the High and Low VVIQ subgroups, and the aphant and hphant individuals. The mean score difference between the Low VVIQ (41.9; SD 11.99) and High VVIQ (75.5; SD 2.72) groups was highly significant: $t(18) = -8.6454$, $p = <0.00001$, 95% confidence interval [−42.2691, −24.9309].

For the effect of gender, the sample of 53 individuals was employed. The effect of gender on VVIQ scores was slight (effect size $d = 0.0034$). A two tailed *t*-test for independent samples showed no statistical difference with t close to zero, t(51) = 0.0123, $p = 0.99$, 95%, confidence interval [−6.9707, 7.0569]. The effect size *d* was 0.0034, indicating a very small effect.

A two tailed *t*-test for independent samples showed that the difference in mean VVIQ scores between the two age groups was not statistically significant, t(18) = −1.0619, $p = 0.302$, 95%, confidence interval [−17.2756, 5.6756]. The effect size *d* of 0.4749 indicated a small effect.

3.2. *Visual Short-Term Visual Memory (VSTM)*

3.2.1. Absolute Error Scores

Four experimental conditions consisted of the presentation of one or three shapes crossed with 1 or 4 s delay between presentation and response. A two-way mixed model ANOVA showed that the difference between the four experimental conditions was statistically significant (F(3, 54) = 65.6892, p = <0.001) (Table S2, Supplementary file) (Figure 4). In all conditions, the Low VVIQ group produced a significantly higher absolute error score than the High VVIQ group: (F(1, 18) = 7.8919, p = 0.012). The interaction between VVIQ group and Condition was not significant (F(3, 54) = 1.9245, p = 0.137). However, the absolute error scores indicated a 38 percent lower error score for the high vividness group. The significant main effect for VVIQ group supports H1.

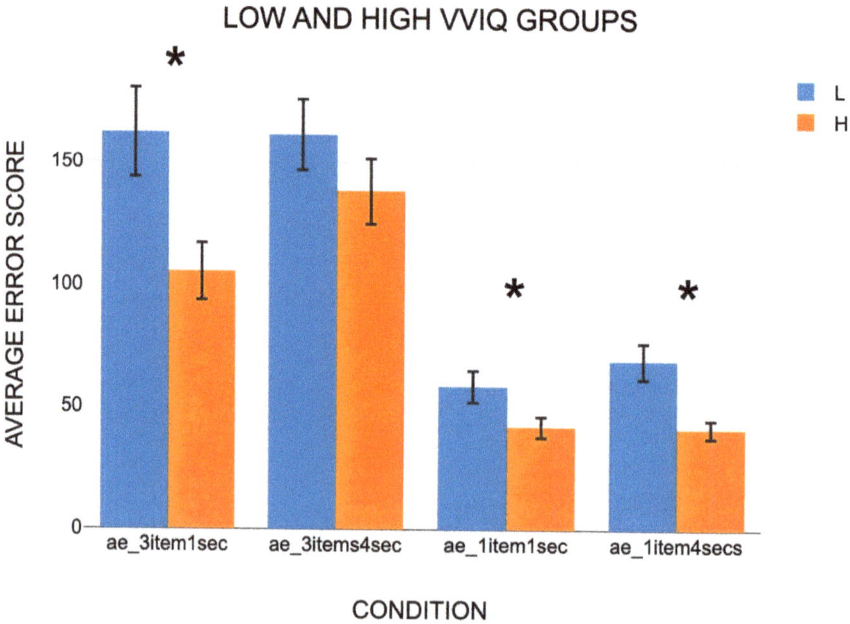

Figure 4. The average error scores for the Low and High VVIQ groups across the four experimental conditions of presentation of one or three shapes crossed with a 1 or 4 s delay between presentation and response. Vertical bars show one standard error. * indicates $p < 0.05$.

A two-way mixed model ANOVA with SIENAX scaling factor as a covariate showed a significant difference between females and males, F(1, 15) = 6.225, p = 0.025, with no interaction between gender and experimental condition, F(3, 45) = 2.457, p = 0.075 (Table S3, Supplementary file). This outcome supports H2.

A two-way ANOVA showed that age was significant F(1, 18) = 6.4309, p = 0.021, with no interaction between condition and age, F(3, 54) = 0.0814, p = 0.97 (Table S4, Supplementary file). This outcome supports H3.

3.2.2. Mean Guessing Rates

Guessing indicates weaker memory. The High VVIQ group had lower guessing rates (M = 0.123, SD = 0.0462) than the Low VVIQ group (M = 0.18, SD = 0.0506), supporting H1 (Table S5, Supplementary file). The high vividness group had a guessing rate that was 43 percent lower than the low vividness group. A mixed model ANOVA showed that the

VVIQ group difference was highly significant: F(1,18) = 11.9845, $p = 0.003$. The interaction with Condition was not significant: F(3,54) = 1.7251, $p = 0.173$. This finding supports H1.

A mixed model ANOVA showed no significant main effect between females and males, F(1, 51) = 0.7, $p = 0.407$, but there was a significant interaction, F(3, 153) = 3.2616, $p = 0.023$. In the two 3 s conditions, males had higher guessing rates than females, supporting H2 (Table S6, Supplementary file).

The younger group had lower guessing rates than the older group. A mixed model ANOVA showed that there was a significant difference between younger and older groups, F(1, 18) = 7.2431, $p = 0.015$, and a significant interaction, F(3, 54) = 4.3263, $p = 0.008$, supporting H3.

3.2.3. Mean Misbinding Rates

The two misbinding conditions that produced scores (misb_3item1sec, misb_3items4sec) were analyzed in a two-way ANOVA with VVIQ group as the second factor. The High VVIQ group had significantly lower misbinding scores ($M = 0.171$, $SD = 0.058$) than the Low VVIQ group ($M = 0.239$, $SD = 0.0816$): F(1,18) = 4.5083, $p = 0.048$. There was a high/low vividness group difference on the misbinding rate of 40 percent. The interaction with Condition was not significant: F(1,18) = 3.2599. $p = 0.088$. The results support H1 (Table S7, Supplementary file).

Combining the VSTM outcomes for average error ($p = 0.012$), guessing rate ($p = 0.003$), and misbinding ($p = 0.048$) into a single p value using Fisher's method gives a p value of 2.38×10^{-5}.

Males made significantly more misbinding errors than females, F(1, 51) = 4.0133, $p = 0.05$, with no interaction between gender and condition, F(1, 51) = 0.0083, $p = 0.928$, supporting H2.

There was no significant difference between younger and older groups, F(1, 18) = 1.8398, $p = 0.192$, and no significant interaction between age and condition, F(1, 18) = 0.0674, $p = 0.798$, offering no support for H3.

3.2.4. Correct Mean Response Times and Proportions

Neither measure produced a significant effect. Gender and age also were not significant factors for either measure; there were no significant interactions.

3.2.5. Aphant and Hphant VSTM Results

Point and interval estimates of effect sizes for the aphant and hphant individuals were calculated using a Bayesian approach [77] (SingleBayes_ES.EXE). The results for the aphant and hphant were compared to those of the main sample and the two VVIQ groups (Table S1, Supplementary file). For 9/18 memory measures, the aphant showed a significantly worse memory than the average of the remaining 52 sample members. Contrary to H10, the aphant performed more strongly than the Low VVIQ group, although not significantly so. The hphant's performance fell in line with the High VVIQ group, and was not superior to the main sample mean, contrary to H11. Although in the predicted direction, the aphant and hphant VSTM scores were not significantly different, so it did not support H12.

3.3. *MRI Volume Scores*

The 57 brain regional volumes were divided into four Sets (Table 2).

Set A consisted of 11 volume measures of four relatively large areas (at least 3000 mm^3).

Set B consisted of 36 hippocampal subfields, 12 each for the left, right, and bilateral volumes.

Set C consisted of three volumes for the left, right, and bilateral amygdala.

Set D consisted of seven volumes for motor areas.

Sets A, B, and D were each subjected to a mixed model four-way ANOVA (IBM SPSS Version 30, Type III sum of squares) with repeated measures on Side (left, right, bilateral) and Region (12 subfields) and two independent groups factors: High vs. Low VVIQ groups and Gender. For Set C, consisting of the amygdala alone, a mixed model three-way ANOVA was conducted with repeated measures on Side (left, right, bilateral) and two independent groups factors: High vs. Low VVIQ groups and Gender.

Table 2. The composition of the dataset of MRI volume scores.

Set	Left	Right	Bilateral
A: 11 volumes	Hippocampus	Hippocampus	Hippocampus
	Whole hippocampus	Whole hippocampus	-
	Area V1	Area V1	Area V1
	Fusiform gyrus	Fusiform gyrus	Fusiform gyrus
B: 36 volumes	Parasubiculum	Parasubiculum	Parasubiculum
	Presubiculum	Presubiculum	Presubiculum
	Subiculum	Subiculum	Subiculum
	CA1	CA1	CA1
	CA3	CA3	CA3
	CA4	CA4	CA4
	GC-ML-DG	GC-ML-DG	GC-ML-DG
	HATA	HATA	HATA
	Fimbria	Fimbria	Fimbria
	Molecular layer	Molecular layer	Molecular layer
	Hippocampal fissure	Hippocampal fissure	Hippocampal fissure
	Hippocampal tail	Hippocampal tail	Hippocampal tail
C: 3 volumes	Amygdala	Amygdala	Amygdala
D: 7 volumes	BA4a	BA4a	-
	BA4p	BA4p	-
	B4	B4	B4

For all four sets, the Age factor involved different samples than the VVIQ group factor and was tested in separate analyses.

3.3.1. Findings for Set A Areas

Cortical regional volumes were significantly higher for the High VVIQ group than for the Low VVIQ group by an average of 9.3 percent (Figure 5). An ANOVA (Table S8, Supplementary file) showed that VVIQ group was significant, $F(1,16) = 7.977$, $p = 0.012$, and also Gender, $F(1,16) = 5.980$, $p = 0.026$. VVIQ Group*Gender was not significant: $F(1,16) = 0.219$, $p = 0.646$. Region was significant, $F(2, 32) = 237.546$, $p < 0.001$, as was Side, $F(2,32) = 1823.167$, $p < 0.001$, and also Side*Region, $F(4,64) = 169.946$, $p < 0.001$. The only significant interaction between VVIQ group or Gender with Region or Side was Side*VVIQ group: $F(2,32) = 4.769$, $p = 0.015$. These results support H4 but not H5.

A mixed model two-way ANOVA for Set A regions with Age and Regions as factors found that neither Age, $F(1, 18) = 0.1726$, $p = 0.683$, nor Age*Region were significant, $F(10, 180) = 0.1726$, $p = 0.683$, offering no support for H6.

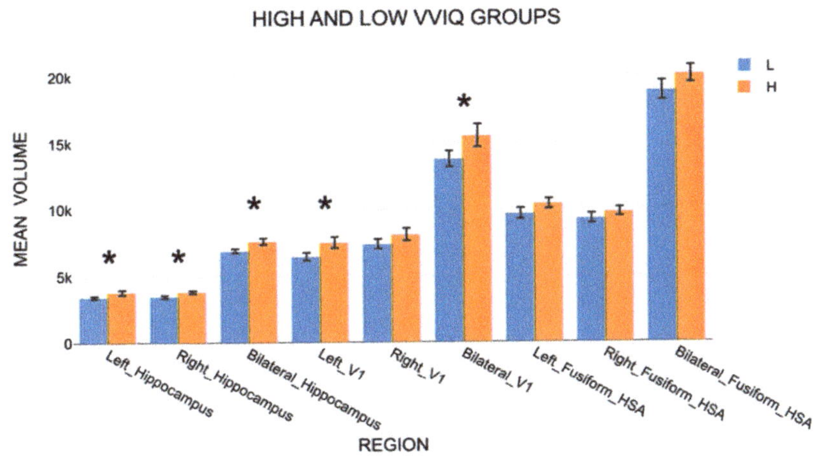

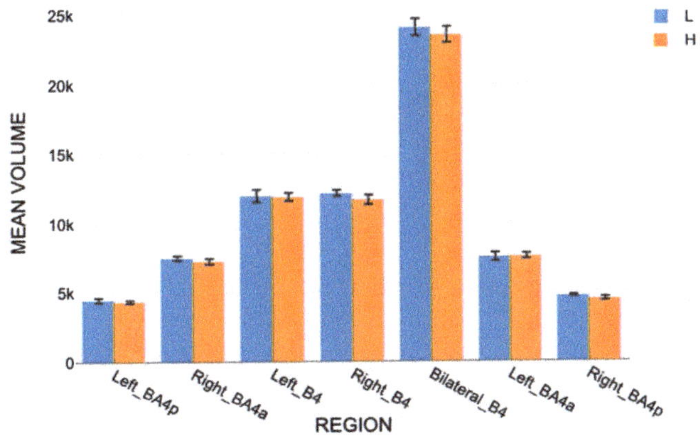

Figure 5. (**Upper panel**) Set A mean volumes for Low and High VVIQ groups. (**Lower panel**) Set D mean volumes for Low and High VVIQ groups. Vertical bars show one standard error. * indicates $p < 0.05$.

3.3.2. Findings for Set B Subfields

A four-way mixed model ANOVA was conducted with VVIQ group and Gender as between groups factors, and Region and Side as within-subject factors (Table S9, Supplementary file). VVIQ group was significant, $F(1,16) = 8.233$, $p = 0.011$ (Partial Eta Squared 0.340), as was Gender, $F(1,16) = 6.252$, $p = 0.024$ (Partial Eta Squared 0.281), but VVIQ group by Gender was not significant: $F(1,16) = 0.156$, $p = 0.698$ (Partial Eta Squared 0.010). Region was statistically significant, $F(11,160) = 834.746$, $p < 0.001$ (Partial Eta Squared 0.981), as was Side, $F(2,32) = 3131.597$, $p < 0.001$ (Partial Eta Squared 0.995), Region*Side: $F(22,352) = 464.562$, $p < 0.001$, and Region*Side*VVIQ group, $F(22,352) = 1.622$, $p = 0.039$. Side*VVIQ group was significant, $F(2,32) = 6.282$, $p = 0.005$, as was Side*Gender, $F(2,32) = 4.807$, $p = 0.015$, but Side*Gender*VVIQ group was not significant: $F(2,32) = 0.151$, $p = 0.861$. Region*VVIQ group was significant, $F(11,352) = 2.112$, $p = 0.022$, as was Region*Gender, $F(11,352) = 1.968$, $p = 0.034$. The mean volumes for the 12 subfields for both genders are shown in Figure 6. These findings support H4.

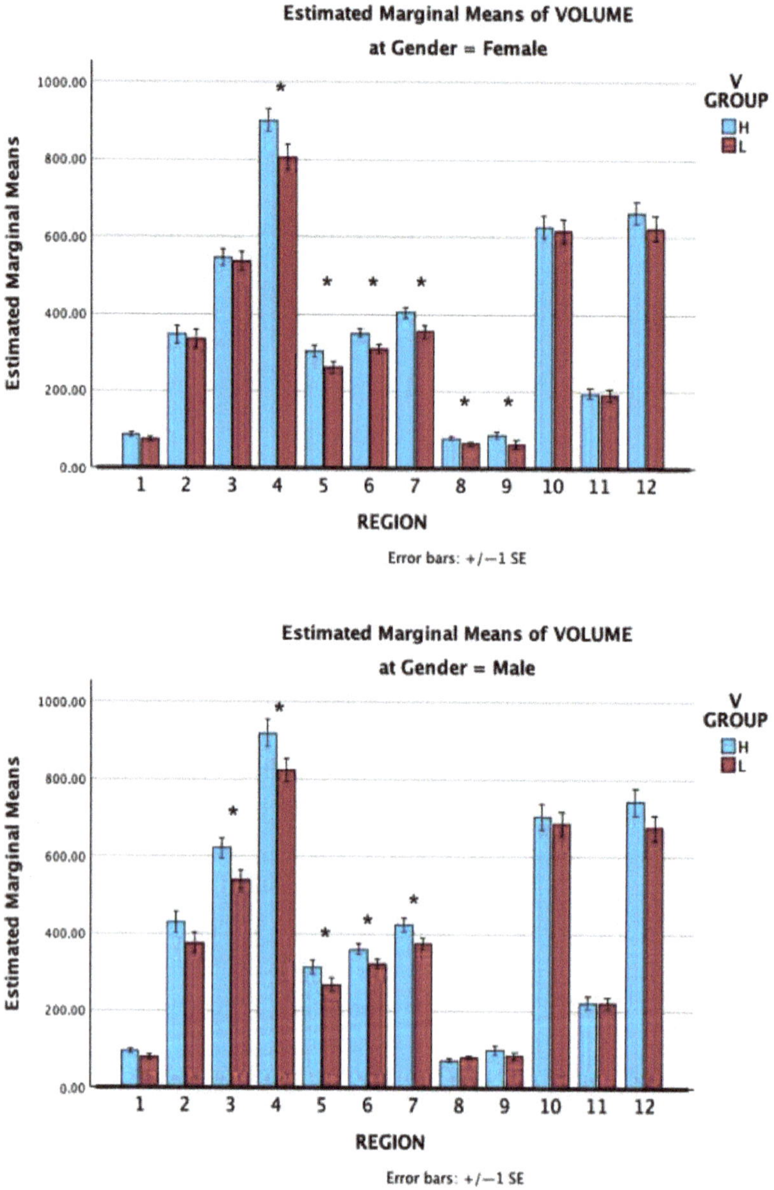

Figure 6. Mean hippocampal subfield volumes for Low and High VVIQ groups: upper chart, females; lower chart, males. Subfields 1–12 are as follows: 1, parasubiculum; 2, presubiculum; 3, subiculum; 4, CA1; 5, CA3; 6, CA4; 7, GC-ML-DG; 8, HATA; 9, fimbria; 10, molecular layer; 11, hippocampal fissure; 12, hippocampal tail. Vertical bars show one standard error. * indicates $p < 0.05$.

Sets A and B in combination contain 47 brain subregions of which 23 showed significantly higher volumes in the High than in the Low VVIQ group (Table S1, Supplementary file). The most significant were in the left hemisphere, left V1 ($p = 0.0284$), left whole hippocampus ($p = 0.0089$), left CA1 ($p = 0.0013$), left CA3 ($p = 0.008$), left CA4 ($p = 0.0078$), and the left hippocampal tail ($p = 0.0455$), all supporting H4.

Contrary to H5, male volumes for Set B exceeded female volumes. A two-way ANOVA with Age between groups and Region as a repeated measures factor showed that Region was significant, F(32, 18) = 783.606, p = < 0.001, but not Age, F(1,18) = 0.0109, p = 0.918, or Age*Region, F(32, 576) = 0.1384, p = 1, contrary to H6.

3.3.3. Findings for the Amygdala

The effect of VVIQ group was not significant (Table S10, Supplementary file). The only significant effect was the Side*Gender interaction: F(2,32) = 4.135, p = 0.025. Females had significantly smaller amygdalae on the left side. A separate ANOVA found no significant Age difference and no Age*Side interaction, supporting H7–H9.

3.3.4. Findings for Set D Subfields

Set D data were analyzed in a four-way ANOVA with the same independent variables as for Sets A and B with two levels of Side (left and right), as bilateral scores were unavailable for BA4a and BA4p (Table S11, Supplementary file). Region was significant, F(2, 32) = 2611.723, p = < 0.001, but not Side, F(1,16) = 0.051, p = 0.824. However, Region*Side showed significance: F(2,32) = 7.457, p = 0.002. Neither VVIQ group nor Gender were significant, nor was the interaction between them. The mean volumes in Set D for the two VVIQ groups were almost identical for all seven regions (Figure 5) in support of H7–H9. There was a significant Region*Gender interaction. In a separate analysis, there was no significant difference between Age groups, F(1, 18) = 0.0035, p = 0.954, and no interactions with Age, F(5, 90) = 0.1373, p = 0.983.

3.3.5. Comparing VMI Active and Control Regions

The study was designed to make it possible to compare the results for regions understood to be "VMI active" with control sites such as motor regions and the amygdala that are understood to be "VMI inactive". Table 3 presents a 2 × 2 breakdown of significant and non-significant differences obtained for the High vs. Low VVIQ group comparisons between active sites in Sets A and B and control sites in Sets C and D.

Table 3. Comparison of significant VVIQ group differences obtained for VMI active and VMI inactive brain regions.

	Sets A and B VMI Active	Sets C and D VMI Inactive	Totals
Significant VVIQ Group difference	24	0	24
Non-significant VVIQ Group difference	23	10	33
Totals	47	10	57

The 2 × 2 chi-squared test gave a chi-squared value (df = 1) of 8.82. $p < 0.005$ (Yates' p value = 0.00886). The comparison shows that 24 significant VVIQ group differences occurred in VMI active Sets A and B only and none occurred in control Sets C and D, as hypothesized.

3.3.6. Findings for Aphant and Hphant

The aphant was aged 74 and the hphant was 60. For 10 of the 47 areas in Sets A and B, the aphant yielded significantly smaller volumes than the sample average (Table 1, column 9). Also, across 47 regions in Sets A and B, the aphant yielded lower volumes

(M = 1793, SD = 2281) than the hphant (M = 2172, SD = 2983). The hphant's volumes exceeded those of the aphant by an average of 37 percent. A paired *t*-test showed that the difference was statistically significant, $t(42) = -5.9735$, $p = < 0.001$, 95% confidence interval [−225.9925, −103.2066], indicating a large effect size (d = 1.80). For only one hippocampal subfield—the left parasubiculum—the hphant yielded a smaller volume than the aphant. These findings support H15.

The aphant's left fimbria volume of 25.23 was 1.73 standard deviations below the group average (66.67), while her right fimbria was even smaller, 10.34, 2.07 standard deviations below the group average (51.55). The aphant also had a small right amygdala of 771.01, 2.04 standard deviations below the group mean of 1310 mm^3. The results are supportive of H13 but not at a statistically significant level.

The findings for the hphant were less significant overall. However, for the left hippocampal tail and the right fusiform gyrus, the hphant's volume significantly exceeded that of the main sample ($p = 0.032$ and $p = 0.0197$, respectively), supporting H14. Collectively, the observed aphant/hphant differences support H15. (Figure 7)

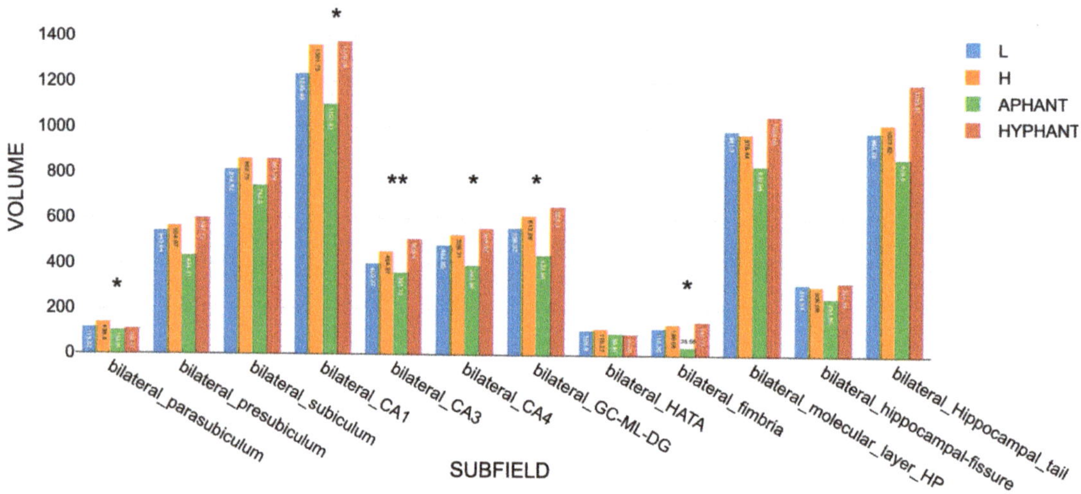

Figure 7. The results for the 12 bilateral hippocampal subfields, which were mirrored by the profiles for leftward and rightward volumes. * indicates $p < 0.05$; ** indicates $p < 0.01$, for the differences between the High and Low VVIQ groups.

Volumes in Set A, including area V1, showed aphant/hphant differences on a similar scale to those in Set B, supporting H15. For the motor areas in Set D, the hphant also had significantly larger volumes than the aphant: $t(6) = -3.767$, $p = 0.009$, d = 1.42.

3.3.7. Other Analyses and Observations

For Sets A and B, there was a notable consistency in the volumes for the High and Low VVIQ groups (Table 1, columns 4 and 6). For 46 of 47 subareas, the High VVIQ had larger volumes than the Low VVIQ group, which has a probability effectively of zero, strongly supporting H4.

Another striking pattern occurred in Sets A and B (Table S1, Supplementary file, marked *). For 30 of the 57 brain regions, the mean volumes were in the predicted sequence: Hphant largest, High VVIQ group second largest, entire sample third, Low VVIQ group fourth, aphant smallest. The chance probability of this result was computed using the

Binomial Theorem. There are 5! = 120 different orderings of these five categories; each occurrence of the specified ordering is a 1-in-120 event (p = 0.008333). In the binomial distribution, the probability of at least k successes in n trials with a probability p of success in any trial is given by: $P(x = k) = (nk)(p)k(1 - p)n - k$. With $n = 57$, $k = 30$, and $p = 0.008333$, the probability of obtaining at least 30 successful matches is effectively zero, providing strong support for H16.

4. Discussion

4.1. Summary of Findings

Table 4 summarises the 16 hypotheses and the study outcomes.

Table 4. Summary of hypotheses and outcomes of investigation.

Label	Hypothesis	Outcome
H1	People with vivid visual imagery have greater visual short-term memory capacity than people with non-vivid visual imagery.	Strongly supported $p < 0.001$
H2	Females have greater visual short-term memory capacity than males.	Supported $p = 0.025$
H3	Younger people have greater visual short-term memory than older people.	Supported $p = 0.01$
H4	In VMIF brain regions, the High VVIQ group have larger volumes than the Low VVIQ group.	Strongly supported $p = 0.012$, $p = 0.011$, $p = 0$ (rank order correct for 46/47 areas in Sets A and B)
H5	In VMI functioning (VMIF) brain regions, females have larger volumes than males.	Unsupported
H6	In VMIF brain areas, younger people have larger volumes than older people.	Unsupported
H7	In non-VMIF brain areas, High and Low VVIQ groups have no volume differences.	Supported
H8	In non-VMIF brain areas, females and males have no volume differences.	Supported
H9	In non-VMIF brain areas, younger and older people have no volume differences.	Supported
H10	Aphant has weaker than average VSTM.	NS
H11	Hphant has stronger than average VSTM.	NS
H12	Hphant has stronger VSTM than aphant.	NS
H13	Aphant has smaller than average VMIF volumes.	Supported in 10 regions
H14	Hphant has larger than average VMIF volumes.	Supported in 2 regions
H15	Hphant has larger VMIF volumes than aphant.	Supported $p < 0.001$
H16	Volume sizes will follow a predictable sequence: hphant, High VVIQ group, entire sample mean, Low VVIQ group, aphant.	Predicted order correct 30/57 times: $p < 0.001$

4.2. General Significance of Findings

In spite of the subjective nature of VVIQ ratings, it is notable that the predicted statistically significant differences have been found in objective measures of VSTM and brain anatomy between High and Low VVIQ scoring participants. Notably, significant VVIQ group differences only occurred in the VMI functionally active brain regions (Sets A and B) and not in the control regions (Sets C and D). The findings are extended to include significantly smaller brain region volumes between an aphantasic individual and the participant group as a whole. Pending further studies, the observed brain region volume differences could provide a helpful signpost to our understanding of VMI vividness differences within the VVIQ distribution.

4.2.1. VSTM

The VSTM findings from this re-analysis of the Tabi et al. [74] dataset are substantially different from theirs. Highly significant differences between Low and High VVIQ groups on three parameters of VSTM performance support H1 that people with vivid visual imagery have greater VSTM capacity than people with non-vivid visual imagery. This result shows a substantial benefit of high VMI vividness to VSTM. The between-group differences were 38%, 43%, and 40% for absolute error, guessing, and misbinding, respectively. These three outcomes had a combined p value of 2.38×10^{-5}. The higher VSTM scores obtained by females and younger participants, respectively, supported H2 that females' VSTM would be stronger, and H3 that younger subjects' VSTM would be stronger than the VSTM of their counterparts.

4.2.2. Brain Regional Differences

The only previous study to investigate brain regional volume differences among individuals differing in VVIQ scores was by Tabi et al. [74]. Their findings of a strong association between VVIQ scores and multiple specific regional brain volumes has been fully corroborated. For the hippocampi and left V1 area, vivid visual imagers had volumes that were 16 percent larger than those for the non-vivid visual imagers. For the right V1 area, a difference of 9% failed to reach significance. The result for the left fusiform gyrus was in the predicted direction with a 7 percent difference, but it missed significance on a one-tailed test ($p = 0.099$). It is likely that larger samples would have yielded significant effects both for the right V1 area and the left fusiform gyrus.

Given the superior VSTM of females, H5 states that females would have larger hippocampal and area V1 volumes, but this was not the case—male volumes exceeded female volumes by an average of around 10 percent. This finding could have several explanations. Firstly, H5 is incorrect, and males have larger volumes than females. Second, volume is only one measure of cortical subfield size along with, for example, area and thickness. What the females lacked in volume may have been compensated for by cortical thickness, which was not measured in this study. Third, it is understood that interconnectivity between areas is an important determinant of VMI, but this was not measured. Fourth, differences in hippocampal volume depend to some degree on the system used to correct volumes for head size [78]. The SIENAX system in this study may have been imperfect for correcting volumes for gender-based head size. Fifthly, our sample sizes were small with only 11 females and 9 males, which is a major limitation.

The age differences were not significant in any volume data. The two age samples were already rather old and the age difference between the two age-based samples was relatively low, with an age gap of only 14 years (Figure 3). Evidently, this age difference was insufficient to observe brain volume differences, and/or the samples were too small, although the age effect was significant for VSTM performance, nevertheless.

4.2.3. The Aphant Profile

This is the first scientific report on anatomical differences in brain region sizes of aphantasic and hyperphantasic individuals. The results show that the aphant yielded significantly smaller volumes than the entire (N = 53) sample average and had 10 significantly smaller volumes than the hphant, five of which were on the right side. The aphant had particularly small fimbria, especially on the right side. It has been reported [79] that fimbria/fornix volume is associated with spatial memory. For this sample of 53 subjects, Pearson's r between VVIQ scores and right fimbria volumes was 0.367 ($p < 0.01$). If it had been possible, it would have been interesting to check the aphant's olfactory sensitivity, which is also claimed to be associated with fornix volume [79]. In spite of her low volumes, the aphant scored rather well on the VSTM task with lower absolute error, guessing, and misbinding scores than the overall group means, although missing statistical significance (Table S1, Supplementary file). One may speculate that over a life of 74 years without voluntary VMI, this subject likely had developed proficient strategies for encoding memory of locations, colours, and shapes usually parts of VMI by using tags, lists, subjective probability estimates, and other symbolic processes. One may expect aphants, routinely and without effort, to be able to tag visual locations mnemonically using quick and direct labels such as "far left", "far left and above", "middle right, straight head", "middle right, down a bit", etc., along with compensatory verbal abilities and strategies deployed as "verbal scaffolding" [80].

It was notable also that the aphant's right amygdala was considerably smaller than the group average. Together with her minute hippocampi and fimbria, one wonders whether the differences collectively contributed to her negligible capacity for VMI? As only a single aphantasic was included, caution is required when interpreting the statistical findings. Further investigation of the neurological signature of aphantasia among large samples should soon provide answers to this question.

4.2.4. The Hphant Profile

The hphant's short-term memory scores were typical of the High VVIQ group as a whole. Although 37 percent higher than the aphant's VSTM scores, the VSTM difference was not statistically significant. Two subfields—left hippocampal tail and the right fusiform gyrus—yielded higher volumes than the entire group average and five volumes were significantly bigger than the aphant's. The overall high VVIQ scores of the High VVIQ sample meant that the entire group was essentially hyperphantasic; their mean VVIQ score was 75.5 with a narrow SD of 2.72. Again, a more in-depth study of a larger sample is necessary.

4.2.5. Consistency with Earlier EEG Findings

Prior EEG research is cited in the Introduction as implicating the left parieto-occipital cortex in mental imagery [30–32], which aligns with a few clinical studies [28,33,81]. This finding is reflected in the current volumetric results and appears consistent with a recent review of hemispheric asymmetries in VMI [82]. In line with {30–32}, the review [82] noted that VMI is associated with frontoparietal networks in both hemispheres; it also suggested lateralization in the temporal lobes favouring the left inferior temporal lobe for endogenous, but not electro-modulated, images.

The present finding of larger V1 volumes in High VVIQ subjects together with their larger hippocampal volumes, particularly on the left side, suggests the possibility of lateral specialization of brain function for VMI. However, contrary to that view is the lack of significance in VVIQ group differences for the left fusiform gyrus volumes. Hippocampal and brain asymmetries in VMI functioning remains a topic for further investigation.

4.2.6. Relationship Between Hippocampus and Area V1

Connectivity in networks between areas acting as nodes is clearly crucial to VMI, although connectivity measures were not included in this study. Area V1 connects directly with the visual areas V2 and V4 and inferior temporal (IT) region that connect with the perirhinal and parahippocampal cortex and directly onwards via the perforant path to the posterior and anterior hippocampus. Tullo et al. (2022) [44] found that the occipito-medial region has an inhibitory influence on temporal regions and an excitatory influence of more anterior on more medial and posterior brain regions. VMI vividness was found to be associated with the connection strength from an occipital area to a parahippocampal area, especially in the left hemisphere. Kvanne et al. (2024) [45] report increased efficiency and clustering in left inferior temporal regions in individuals with more vivid visual imagery. Thus, area V1, the left IT region, and the anterior and posterior hippocampus all appear crucial to VMI vividness. In this study, the left area V1 volume and the bilateral V1 volume were larger in the High VVIQ group than in the Low VVIQ group. This finding is consistent with those of Dalton et al., 2022 [83] who used track-density imaging to examine anatomical connectivity between the cortex and the anterior–posterior axis of the in vivo human hippocampus. Their findings indicate the possibility of two medial hippocampal hubs having high anatomical connectivity: a posterior medial hub linked with visuospatial processing areas in medial parietal and occipital cortices and an anterior medial hub linked with temporopolar, inferotemporal, medial parietal, and occipital areas.

4.3. *Strengths and Limitations*

4.3.1. Independence and Transparency

The data for this study were collected and curated independently of this author at the University of Oxford Medical Sciences Division, Oxford, United Kingdom. The data are shared at the Open Science Foundation website. The author had no role in the data collection or the recruitment of subjects, which were carried out entirely independently and transparently. All study data are downloadable from a public repository: https://osf.io/q37vn/ accessed on 26 June 2025.

4.3.2. Measures Used

Volume is only one of several subfield size measures including also area and thickness. The females' volume lack may have been compensated by thickness, which was not measured here. Interconnectivity between areas is another possible determinant of vivid VMI but this was also not measured. Hippocampal subfield measures depend to some degree on the correction system used to calibrate measures against head size [78]. The reliability of the SIENAX system in this study for correcting volumes for gender-based head size is an unknown, possibly confounding, factor.

A reviewer mentioned, given concerns about the accuracy of Freesurfer-derived subfields, that it would ideally have been important to clarify whether, for the original data, the authors had conducted visual inspections or carried out other quality control of the Freesurfer outputs. Unfortunately, details of error-correction with the Freesurfer-derived subfield volume scores were unavailable.

4.3.3. Small Group Sizes

It has been stated that the "extreme groups" protocol is more sensitive than correlation in detecting cognitive differences between subjects varying in VMI vividness. The study protocol used here revealed a large difference in VSTM between High and Low VVIQ groups in spite of the small Ns.

The downside of the "extreme groups" protocol is the reduction in sample size from 53 to 20. Fifty-three is already not large, but the reduction to twenty, is much worse for statistical power. Substantial VVIQ differences in volumes could have overcome the relative lack of power of the small group Ns, but in the case of gender and age, the study was inconclusive in these regards.

As noted, the study included only one aphantasic and one hyperphantasic subject, and so considerable caution is warranted in interpreting the statistical findings about them.

5. Conclusions

(1) Are VMI-vividness differences measured with the VVIQ actually measuring anything? If we are enabled to make rich, precise, falsifiable predictions using VVIQ measures, then the answer is affirmative. VVIQ differences are measuring genuine VMI differences, and, in this study as in many others, they have produced a set of highly significant and meaningful findings.

(2) What functions do the vividness differences serve? High VMI-vividness serves three primary functions: (A) remembering recent and distant-past stimuli, scenarios, episodes, and events; (B) anticipating, foreseeing, and simulating near and distant future stimuli, scenarios, episodes, and events; and, (C) constructing phantasy for dreams, imaginary stimuli, scenarios, episodes, and events. Low or absent VMI require alternative, non-imagistic mnemonic strategies, such as scaffolding, tagging, and listing, to perform tasks but, in some cases, less efficiently.

(3) What is their neurological foundation? VMI-vividness differences are founded, at least in part, on a highly varied set of regional brain volumes which show significant, systematic associations with VVIQ scores.

Supplementary Materials: The following supporting information can be downloaded at: https://www.mdpi.com/article/10.3390/vision9030053/s1, Table S1: VVIQ scores, STVM parameters and MRI indices of subfield cortical volume for the high VVIQ group, low VVIQ group, overall sample means, hyperphantasic and aphantasic individuals; Table S2: Two-way mixed model ANOVA with VVIQ group as a between groups factor, and Condition as a repeated measures factor. Dependent variable: Absolute Error scores; Table S3: Two-way mixed model ANOVA with Gender and VVIQ group as between groups factors, and SIENAX scaling factor as a covariate. Dependent variable: Absolute Error scores; Table S4: Two-way mixed model ANOVA with Age Group as a between groups factor, and Condition as a repeated measures factor. Dependent variable: Absolute Error scores; Table S5: Two-way mixed model ANOVA with VVIQ group as a between groups factor, and Condition as a repeated measures factor. Dependent variable: Guessing scores; Table S6: Two-way mixed model ANOVA with Gender as a between groups factor, and Condition as a repeated measures factor. Dependent variable: Guessing scores; Table S7: Two-way mixed model ANOVA with VVIQ group as a between groups factor, and Condition as a repeated measures factor. Dependent variable: Mis-binding rates; Table S8: Four-way mixed model ANOVA with VVIQ group and gender as between groups factors, and region and side as within-subjects factors. Dependent variable: Set A volumes; Table S9: Four-way mixed model ANOVA with VVIQ group and gender as between groups factors, and region and side as within-subjects factors. Dependent variable: Set B volumes; Table S10: Three-way mixed model ANOVA with VVIQ group and gender as between groups factors, and side as within-subjects factors. Dependent variable: Amygdala volumes; Table S11: Four-way mixed model ANOVA with VVIQ group and gender as between groups factors, and region and side as within-subjects factors. Dependent variable: Set D volumes.

Funding: This research received no external funding.

Institutional Review Board Statement: As a re-analysis of existing data, this study did not require ethical approval.

Informed Consent Statement: Informed consent was obtained by the original investigators [74] from all subjects involved in the study.

Data Availability Statement: All data used in this study are available at: https://osf.io/q37vn/ accessed on 26 June 2025.

Conflicts of Interest: The author declares no conflict of interest.

References

1. Marks, D.F. The action cycle theory of perception and mental imagery. *Vision* **2023**, *7*, 12. [CrossRef] [PubMed]
2. Marks, D.F. Individual Differences in the Vividness of Visual Imagery and Their Effect on Function. In *The Function and Nature of Imagery*; Sheehan, P.W., Ed.; Academic Press: New York, NY, USA, 1972; pp. 83–108.
3. Marks, D.F. Phenomenological Studies of Visual Mental Imagery: A Review and Synthesis of Historical Datasets. *Vision* **2023**, *7*, 67. [CrossRef]
4. Bilzer, M.; Monzel, M. The Phenomenology of Offline Perception: Multisensory Profiles of Voluntary Mental Imagery and Dream Imagery. *Vision* **2025**, *9*, 37. [CrossRef]
5. Belfi, A.M. Emotional valence and vividness of imagery predict aesthetic appeal in music. *Psychomusicol. Music. Mind Brain* **2019**, *29*, 128. [CrossRef]
6. Marks, D.F. *The General Theory of Behaviour: A Sourcebook*; Cambridge Scholars Publishing: Newcastle upon Tyne, UK, 2025. Available online: https://www.cambridgescholars.com/product/978-1-0364-4245-3 (accessed on 3 July 2025).
7. Marks, D.F. Visual imagery differences in the recall of pictures. *Br. J. Psychol.* **1973**, *64*, 17–24. [CrossRef] [PubMed]
8. Baddeley, A.D.; Andrade, J. Working memory and the vividness of imagery. *J. Exp. Psychol. Gen.* **2000**, *129*, 126–145. [CrossRef] [PubMed]
9. Chkhaidze, A.; Coulson, S.; Kiyonaga, A. Individual Differences in Preferred Thought Formats Predict Features of Narrative Recall. *Proc. Annu. Meet. Cogn. Sci. Soc.* **2023**, *45*. Available online: https://escholarship.org/uc/item/0jm2d2rm (accessed on 3 July 2025).
10. Keogh, R.; Pearson, J. Mental imagery and visual working memory. *PLoS ONE* **2011**, *6*, e29221. [CrossRef]
11. Keogh, R.; Pearson, J. The sensory strength of voluntary visual imagery predicts visual working memory capacity. *J. Vis.* **2014**, *14*, 1427–1431. [CrossRef]
12. Gur, R.C.; Hilgard, E.R. Visual imagery and the discrimination of differences between altered pictures simultaneously and successively presented. *Br. J. Psychol.* **1975**, *66*, 341–345. [CrossRef]
13. McKelvie, S.J.; Demers, E.G. Individual differences in reported visual imagery and memory performance. *Br. J. Psychol.* **1979**, *70*, 51–57. [CrossRef] [PubMed]
14. Berger, G.H.; Gaunitz, S.C. Self-rated imagery and encoding strategies in visual memory. *Br. J. Psychol.* **1979**, *70*, 21–24. [CrossRef]
15. Slee, J.A. Individual differences in visual imagery ability and the retrieval of visual appearances. *J. Ment. Imag.* **1980**, *4*, 93–113.
16. McKelvie, S.J. The VVIQ as a psychometric test of individual differences in visual imagery vividness: A critical quantitative review and plea for direction. *J. Ment. Imag.* **1995**, *19*, 1–106.
17. Rodway, P.; Gillies, K.; Schepman, A. Vivid imagers are better at detecting salient changes. *J. Individ. Differ.* **2006**, *27*, 218–228. [CrossRef]
18. Jacobs, C.; Schwarzkopf, D.S.; Silvanto, J. Visual working memory performance in aphantasia. *Cortex* **2018**, *105*, 61–73. [CrossRef] [PubMed]
19. Heuer, F.; Fischman, D.; Reisberg, D. Why does vivid imagery hurt colour memory? *Can. J. Psychol./Rev. Can. De Psychol.* **1986**, *40*, 161–175. [CrossRef]
20. Reisberg, D.; Culver, L.C.; Heuer, F.; Fischman, D. Visual memory: When imagery vividness makes a difference. *J. Ment. Imag.* **1986**, *10*, 51–74.
21. Reisberg, D.; Leak, S. Visual imagery and memory for appearance: Does Clark Gable or George C. Scott have bushier eyebrows? *Can. J. Psychol./Rev. Can. De Psychol.* **1987**, *41*, 521–526. [CrossRef]
22. Zhang, Z.; Liu, Y.; Yang, J.; Li, C.; Marks, D.F.; Sala, S.D.; Zhao, B. The Establishment of the Chinese Version of the Vviq (Vviq-C) and the Effects of Age and Gender on the Vividness of Visual Imagery. *J. Psychiatry Psychiatr. Disord.* **2025**, *9*, 163–171. [CrossRef]
23. Zeman, A.; Dewar, M.; Della Sala, S. Lives without imagery—Congenital aphantasia. *Cortex* **2015**, *73*, 378–380. [CrossRef] [PubMed]
24. Zeman, A.; Milton, F.; Della Sala, S.; Dewar, M.; Frayling, T.; Gaddum, J.; Hattersley, A.; Heuerman-Williamson, B.; Jones, K.; MacKisack, M.; et al. Phantasia–The psychological significance of lifelong visual imagery vividness extremes. *Cortex* **2020**, *130*, 426–440. [CrossRef] [PubMed]
25. Blomkvist, A.; Marks, D.F. Defining and 'diagnosing' aphantasia: Condition or individual difference? *Cortex* **2023**, *169*, 220–234. [CrossRef] [PubMed]

26. Jin, F.; Hsu, S.-M.; Li, Y. A Systematic Review of Aphantasia: Concept, Measurement, Neural Basis, and Theory Development. *Vision* **2024**, *8*, 56. [CrossRef]
27. Geschwind, N. Disconnexion syndromes in animals and man. I. *Brain* **1965**, *88*, 237–294. [CrossRef]
28. Farah, M.J.; Levine, D.N.; Calvanio, R. A case study of mental imagery deficit. *Brain Cogn.* **1988**, *8*, 147–164. [CrossRef]
29. Marks, D.F. On the relationship between imagery, body and mind. In *Imagery: Current Developments*; Hampson, P.J., Marks, D.F., Richardson, J.T.E., Eds.; Routledge: London, UK, 1990; pp. 1–38.
30. Marks, D.F.; Uemura, K.; Tatsuno, J.; Imamura, Y.; Ashida, H. Topographical analysis of alpha attenuation in mental imagery and calculation. *Electroencephalogr. Clin. Neurophysiol.* **1985**, *61*, S142. [CrossRef]
31. Marks, D.; Uemura, K.; Tatsuno, J.; Imamura, Y. EEG topographical analysis of imagery. In *Contemporary Psychology: Biological Processes and Theoretical Issues*; Elsevier: Amsterdam, The Netherlands, 1985; pp. 211–223.
32. Marks, D.F.; Isaac, A.R. Topographical distribution of EEG activity accompanying visual and motor imagery in vivid and non-vivid imagers. *Br. J. Psychol.* **1995**, *86*, 271–282. [CrossRef]
33. Farah, M.J. The neurological basis of mental imagery: A componential analysis. *Cognition* **1984**, *18*, 245–272. [CrossRef]
34. Kosslyn, S.M.; Thompson, W.L. When is early visual cortex activated during visual mental imagery? *Psychol. Bull.* **2003**, *129*, 723. [CrossRef]
35. Amedi, A.; Malach, R.; Pascual-Leone, A. Negative BOLD differentiates visual imagery and perception. *Neuron* **2005**, *48*, 859–872. [CrossRef] [PubMed]
36. Cui, X.; Jeter, C.B.; Yang, D.; Montague, P.R.; Eagleman, D.M. Vividness of mental imagery: Individual variability can be measured objectively. *Vis. Res.* **2007**, *47*, 474–478. [CrossRef] [PubMed]
37. Dijkstra, N.; Bosch, S.E.; van Gerven, M.A. Shared neural mechanisms of visual perception and imagery. *Trends Cogn. Sci.* **2019**, *23*, 423–434. [CrossRef]
38. Pearson, J. The human imagination: The cognitive neuroscience of visual mental imagery. *Nat. Rev. Neurosci.* **2019**, *20*, 624–634. [CrossRef] [PubMed]
39. Spagna, A.; Hajhajate, D.; Liu, J.; Bartolomeo, P. Visual mental imagery engages the left fusiform gyrus, but not the early visual cortex: A meta-analysis of neuroimaging evidence. *Neurosci. Biobehav. Rev.* **2021**, *122*, 201–217. [CrossRef]
40. Dijkstra, N. Uncovering the Role of the Early Visual Cortex in Visual Mental Imagery. *Vision* **2024**, *8*, 29. [CrossRef]
41. Genç, E.; Bergmann, J.; Singer, W.; Kohler, A. Surface Area of Early Visual Cortex Predicts Individual Speed of Traveling Waves During Binocular Rivalry. *Cereb. Cortex* **2015**, *25*, 1499–1508. [CrossRef]
42. Fulford, J.; Milton, F.; Salas, D.; Smith, A.; Simler, A.; Winlove, C.; Zeman, A. The neural correlates of visual imagery vividness–An fMRI study and literature review. *Cortex* **2018**, *105*, 26–40. [CrossRef]
43. Duan, S.; Li, Q.; Yang, J.; Yang, Q.; Li, E.; Liu, Y.; Jiang, L.; Li, C.; Zhao, B. Precuneus activation correlates with the vividness of dynamic and static imagery: An fMRI study. *Front. Hum. Neurosci.* **2025**, *19*, 1516058. [CrossRef]
44. Tullo, M.G.; Almgren, H.; Van de Steen, F.; Sulpizio, V.; Marinazzo, D.; Galati, G. Individual differences in mental imagery modulate effective connectivity of scene-selective regions during resting state. *Brain Struct. Funct.* **2022**, *227*, 1831–1842. [CrossRef]
45. Kvamme, T.L.; Lumaca, M.; Zana, B.; Paunovic, D.; Silvanto, J.; Sandberg, K. Vividness of Visual Imagery Supported by Intrinsic Structural-Functional Brain Network Dynamics. *bioRxiv* **2024**. [CrossRef]
46. Blomkvist, A. Shaping the Space: A Role for the Hippocampus in Mental Imagery Formation. *Vision* **2025**, *9*, 2. [CrossRef]
47. Ellis, H. *Man and Woman: A Study of Secondary Sexual Characters*; Walter Scott, Ltd.: London, UK, 1894.
48. Horgan, T.G.; Mast, M.S.; Hall, J.A.; Carter, J.D. Gender Differences in Memory for the Appearance of Others. *Personal. Soc. Psychol. Bull.* **2004**, *30*, 185–196. [CrossRef]
49. McBurney, D.H.; Gaulin, S.J.; Devineni, T.; Adams, C. Superior spatial memory of women: Stronger evidence for the gathering hypothesis. *Evol. Hum. Behav.* **1997**, *18*, 165–174. [CrossRef]
50. Choi, J.; Silverman, I. Sexual dimorphism in spatial behaviors: Applications to route learning. *Evol. Cogn.* **1996**, *2*, 165–171.
51. James, T.W.; Kimura, D. Sex differences in remembering the locations of objects in an array: Location-shifts versus location-exchanges. *Evol. Hum. Behav.* **1997**, *18*, 155–163. [CrossRef]
52. McGivern, R.F.; Huston, J.P.; Byrd, D.; King, T.; Siegle, G.J.; Reilly, J. Sex differences in visual recognition memory: Support for a sex-related difference in attention in adults and children. *Brain Cogn.* **1997**, *34*, 323–336. [CrossRef]
53. Levy, L.J.; Astur, R.S.; Frick, K.M. Men and women differ in object memory but not performance of a virtual radial maze. *Behav. Neurosci.* **2005**, *119*, 853. [CrossRef]
54. Silverman, I.; Choi, J.; Peters, M. The hunter-gatherer theory of sex differences in spatial abilities: Data from 40 countries. *Arch. Sex. Behav.* **2007**, *36*, 261–268. [CrossRef]
55. Russell, E.M.; Longstaff, M.G.; Winskel, H. Sex differences in eyewitness memory: A scoping review. *Psychon. Bull. Rev.* **2024**, *31*, 985–999. [CrossRef] [PubMed]

56. Herlitz, A.; Rehnman, J. Sex differences in episodic memory. *Curr. Dir. Psychol. Sci.* **2008**, *17*, 52–56. [CrossRef]
57. Asperholm, M.; Hogman, N.; Rafi, J.; Herlitz, A. What did you do yesterday? A metaanalysis of sex differences in episodic memory. *Psychol. Bull.* **2019**, *145*, 785–821. [CrossRef]
58. Voyer, D.; Voyer, S.D.; Saint-Aubin, J. Sex differences in visual-spatial working memory: A meta-analysis. *Psychon. Bull. Rev.* **2017**, *24*, 307–334. [CrossRef]
59. Filipek, P.A.; Richelme, C.; Kennedy, D.N.; Caviness, V.S., Jr. The young adult human brain: An MRI-based morphometric analysis. *Cereb. Cortex* **1994**, *4*, 344–360. [CrossRef]
60. Filipek, P.A.; Semrud-Clikeman, M.; Steingard, R.J.; Renshaw, P.F.; Kennedy, D.N.; Biederman, J. Volumetric MRI analysis comparing subjects having attention-deficit hyperactivity disorder with normal controls. *Neurology* **1997**, *48*, 589–601. [CrossRef]
61. Murphy, D.G.M.; DeCarli, C.; Mclntosh, A.R.; Daly, E.; Mentis, M.J.; Pietrini, P.; Szczepanik, J.; Schapiro, M.B.; Grady, C.L.; Horwitz, B.; et al. Sex differences in human brain morphometry and metabolism: An in vivo quantitative magnetic resonance imaging and positron emission tomography study on the effect of aging. *Arch. Gen. Psychiatry* **1996**, *53*, 585–594. [CrossRef]
62. Paus, T.; Otaky, N.; Caramanos, Z.; Macdonald, D.; Zijdenbos, A.; D'Avirro, D.; Gutmans, D.; Holmes, C.; Tomaiuolo, F.; Evans, A.C. In vivo morphometry of the intrasulcal gray matter in the human cingulate, paracingulate, and superior-rostral sulci: Hemispheric asymmetries, gender differences and probability maps. *J. Comp. Neurol.* **1996**, *376*, 664–673. [CrossRef]
63. Schlaepfer, T.E.; Harris, G.J.; Tien, A.Y.; Peng, L.; Lee, S.; Pearlson, G.D. Structural differences in the cerebral cortex of healthy female and male subjects: A magnetic resonance imaging study. *Psychiatry Res. Neuroimaging* **1995**, *61*, 129–135. [CrossRef]
64. Guadalupe, T.; Zwiers, M.P.; Wittfeld, K.; Teumer, A.; Vasquez, A.A.; Hoogman, M.; Hagoort, P.; Fernandez, G.; Buitelaar, J.; van Bokhoven, H.; et al. Asymmetry within and around the human planum temporale is sexually dimorphic and influenced by genes involved in steroid hormone receptor activity. *Cortex* **2015**, *62*, 41–55. [CrossRef]
65. Witelson, S.F.; Glezer, I.I.; Kigar, D.L. Women have greater density of neurons in posterior temporal cortex. *J. Neurosci.* **1995**, *15*, 3418–3428. [CrossRef] [PubMed]
66. Giedd, J.N.; Snell, J.W.; Lange, N.; Rajapakse, J.C.; Casey, B.J.; Kozuch, P.L.; Vaituzis, A.C.; Vauss, Y.C.; Hamburger, S.D.; Kaysen, D.; et al. Quantitative magnetic resonance imaging of human brain development: Ages 4–18. *Cereb. Cortex* **1996**, *6*, 551–560. [CrossRef]
67. Persson, J.; Spreng, R.N.; Turner, G.; Herlitz, A.; Morell, A.; Stening, E.; Wahlund, L.O.; Wikström, J.; Söderlund, H. Sex differences in volume and structural covariance of the anterior and posterior hippocampus. *Neuroimage* **2014**, *99*, 215–225. [CrossRef]
68. Dalton, M.A.; Zeidman, P.; McCormick, C.; Maguire, E.A. Differentiable processing of objects, associations, and scenes within the hippocampus. *J. Neurosci.* **2018**, *38*, 8146–8159. [CrossRef]
69. Fjell, A.M.; Sneve, M.H.; Amlien, I.K.; Grydeland, H.; Mowinckel, A.M.; Vidal-Piñeiro, D.; Sørensen, Ø.; Walhovd, K.B. Stable hippocampal correlates of high episodic memory function across adulthood. *Sci. Rep.* **2025**, *15*, 8816. [CrossRef]
70. Harrison, T.M.; Weintraub, S.; Mesulam, M.M.; Rogalski, E. Superior memory and higher cortical volumes in unusually successful cognitive aging. *J. Int. Neuropsychol. Soc.* **2012**, *18*, 1081–1085. [CrossRef]
71. Dawes, A.J.; Keogh, R.; Robuck, S.; Pearson, J. Memories with a blind mind: Remembering the past and imagining the future with aphantasia. *Cognition* **2022**, *227*, 105192. [CrossRef]
72. Monzel, M.; Leelaarporn, P.; Lutz, T.; Schultz, J.; Brunheim, S.; Reuter, M.; McCormick, C. Hippocampal-occipital connectivity reflects autobiographical memory deficits in aphantasia. *elife* **2024**, *13*, RP94916. [CrossRef]
73. Blomkvist, A. Aphantasia: In search of a theory. *Mind Lang.* **2023**, *38*, 866–888. [CrossRef]
74. Tabi, Y.A.; Maio, M.R.; Attaallah, B.; Dickson, S.; Drew, D.; Idris, M.I.; Kienast, A.; Klar, V.; Nobis, L.; Plant, O.; et al. Vividness of visual imagery questionnaire scores and their relationship to visual short-term memory performance. *Cortex* **2022**, *146*, 186–199. [CrossRef]
75. Pertzov, Y.; Miller, T.D.; Gorgoraptis, N.; Caine, D.; Schott, J.M.; Butler, C.; Husain, M. Binding deficits in memory following medial temporal lobe damage in patients with voltage-gated potassium channel complex antibody-associated limbic encephalitis. *Brain* **2013**, *136*, 2474–2485. [CrossRef]
76. Patenaude, B.; Smith, S.M.; Kennedy, D.N.; Jenkinson, M.A. Bayesian model of shape and appearance for subcortical brain segmentation. *Neuroimage* **2011**, *56*, 907–922. [CrossRef] [PubMed]
77. Crawford, J.R.; Garthwaite, P.H. Comparison of a single case to a control or normative sample in neuropsychology: Development of a Bayesian approach. *Cogn. Neuropsychol.* **2007**, *24*, 343–372. [CrossRef] [PubMed]
78. Perlaki, G.; Orsi, G.; Plozer, E.; Altbacker, A.; Darnai, G.; Nagy, S.A.; Horvath, R.; Toth, A.; Doczi, T.; Kovacs, N.; et al. Are there any gender differences in the hippocampus volume after head-size correction? A volumetric and voxel-based morphometric study. *Neurosci. Lett.* **2014**, *570*, 119–123. [CrossRef] [PubMed]
79. Dahmani, L.; Courcot, B.; Near, J.; Patel, R.; Amaral, R.S.C.; Chakravarty, M.M.; Bohbot, V.D. Fimbria-Fornix Volume Is Associated with Spatial Memory and Olfactory Identification in Humans. *Frontiers Systematic Neuroscience* **2020**, *13*, 87. [CrossRef] [PubMed]

80. Bainbridge, W.A.; Pounder, Z.; Eardley, A.F.; Baker, C.I. Quantifying aphantasia through drawing: Those without visual imagery show deficits in object but not spatial memory. *Cortex* **2021**, *135*, 159–172. [CrossRef]
81. Thorudottir, S.; Sigurdardottir, H.M.; Rice, G.E.; Kerry, S.J.; Robotham, R.J.; Leff, A.P.; Starrfelt, R. The architect who lost the ability to imagine: The cerebral basis of visual imagery. *Brain Sci.* **2020**, *10*, 59. [CrossRef]
82. Liu, J.; Spagna, A.; Bartolomeo, P. Hemispheric asymmetries in visual mental imagery. *Brain Struct. Funct.* **2022**, *227*, 697–708. [CrossRef]
83. Dalton, M.A.; D'Souza, A.; Lv, J.; Calamante, F. New insights into anatomical connectivity along the anterior–posterior axis of the human hippocampus using in vivo quantitative fibre tracking. *eLife* **2022**, *11*, e76143. [CrossRef]

MDPI AG
Grosspeteranlage 5
4052 Basel
Switzerland
Tel.: +41 61 683 77 34

Vision Editorial Office
E-mail: vision@mdpi.com
www.mdpi.com/journal/vision

www.ingramcontent.com/pod-product-compliance
Lightning Source LLC
Chambersburg PA
CBHW061556210926
51850CB00009B/2117
* 9 7 8 3 7 2 5 8 4 7 6 1 7 *